# IMAGES OF WAR
## **CHALLENGER 2**

*This book is dedicated to the memory of Sergeant David 'Bob' Monkhouse, Royal Dragoon Guards; killed in action in Afghanistan in 2010. His kind help in assembling material for this study in its early stages was much appreciated. We also dedicate this work to all British service personnel killed and wounded in Operations Telic and Herrick, and to their families.*

# IMAGES OF WAR
## CHALLENGER 2:
## THE BRITISH MAIN BATTLE TANK

RARE PHOTOGRAPHS FROM WARTIME ARCHIVES

M P ROBINSON AND ROB GRIFFIN

Pen & Sword
**MILITARY**

First published in Great Britain in 2017 by
**PEN & SWORD MILITARY**
an imprint of
Pen & Sword Books Ltd,
47 Church Street, Barnsley,
South Yorkshire.
S70 2AS

Copyright © M P Robinson and Rob Griffin 2017

ISBN 978-1-47389-665-9

The right of M P Robinson and Rob Griffin to be identified as Authors of this work has been asserted by them in accordance with the Copyright, Designs and Patents Act 1988.

A CIP catalogue record for this book is available from the British Library.

*All rights reserved. No part of this book may be reproduced or transmitted in any form or by any means, electronic or mechanical including photocopying, recording or by any information storage and retrieval system, without permission from the Publisher in writing.*

Typeset by Mac Style Ltd, Bridlington, East Yorkshire
Printed and bound in India by Replika Press Pvt. Ltd.

*Pen & Sword Books Ltd incorporates the imprints of*
Pen & Sword Books Limited incorporates the imprints of Atlas, Archaeology, Aviation, Discovery, Family History, Fiction, History, Maritime, Military, Military Classics, Politics, Select, Transport, True Crime, Air World, Frontline Publishing, Leo Cooper, Remember When, Seaforth Publishing, The Praetorian Press, Wharncliffe Local History, Wharncliffe

*For a complete list of Pen & Sword titles please contact:*
PEN & SWORD BOOKS LIMITED
47 Church Street, Barnsley, South Yorkshire, S70 2AS, England.
E-mail: enquiries@pen-and-sword.co.uk
Website: www.pen-and-sword.co.uk

# Contents

| | | |
|---|---|---|
| *Acknowledgements* | | vii |
| *Preface* | | viii |
| *Introduction* | | ix |
| **Chapter 1:** | Chieftain, MBT80 and the *Shir* | 1 |
| **Chapter 2:** | Challenger 1 | 9 |
| **Chapter 3:** | The Vickers Mk.7 | 11 |
| **Chapter 4:** | GSR 4026, CAT 87 and the End of the Cold War | 13 |
| **Chapter 5:** | The 120mm L30A1 gun | 18 |
| **Chapter 6:** | The New Vickers MBT | 20 |
| **Chapter 7:** | Options for Change, Operation Granby and the Selection of the Vickers Challenger 2 | 24 |
| **Chapter 8:** | Challenger 2 Prototypes | 28 |
| **Chapter 9:** | The Challenger 2 Described | 29 |
| **Chapter 10:** | Early Service | 33 |
| **Chapter 11:** | *Saif Sareea II* | 48 |
| **Chapter 12:** | Operation Telic | 54 |
| **Chapter 13:** | Occupation and Urban Combat in Iraq | 68 |
| **Chapter 14:** | Challenger 2 Variants | 83 |
| **Chapter 15:** | The Challenger 2's career since 2003 | 117 |
| **Chapter 16:** | The Future | 136 |
| *Notes* | | 152 |

Men of the SCOTS DG prepare to conduct a live-fire exercise at the Joint Multinational Training Command's Grafenwoehr Training Area in October 2012. The American Exercise Saber Junction included elements of units from 18 different nations to test the different NATO armies capabilities to form an operational coalition force. (*U.S. Army photo by Specialist J. Leonard*)

A Sergeant of the SCOTS DG hands APFSDS rounds up to his crewmen at the Joint Multinational Training Command's Grafenwoehr range area. The range period preceded tactical training during Exercise Saber Junction in October 2012. (*U.S. Army photo by Staff Sergeant J. Salgado*)

# Acknowledgements

We would like to express our sincere thanks to Major-General Sir Laurence New CB, CBE, to Colonel Peter Barry CBE, Lieutenant-Colonel Brian Trueman OBE and Brigadier Johnny Torrens-Spence CBE. Thanks to Dennis Lunn for his extensive help on the subject of the Challenger 2 prototypes and to Tim Neate for allowing us to use selections from his splendid photo collection. We also thank Home Headquarters Royal Dragoon Guards, Home Headquarters Royal Tank Regiment, Stephen May, Simon Dunstan, Dick Taylor, Richard Stewart, Lawrence Skuse, Pierre Delattre, Peter Brown, Peter Breakspear, Tom Coates, Matt Noone, David Stickland, Adarsh Ramamurthy, James Patterson, Anthony Ryan, Colin Rosenwould, P. Drabble, J.-W. DeBoer, Keith Paget, James Patterson, Marylin Suckling Gear and Ron Mihalko for their kind help in obtaining reference materials and photographs for this work.

M.P. Robinson and Rob Griffin

A Challenger 2 of the SCOTS DG deployed in training on snowy ground prior to Exercise Saber Junction in October 2012. The lion rampant of Scotland is prominently stencilled on the turret's side. (*U.S. Army photo by B. Fletcher*)

# Preface

This book describes the history of the Challenger 2 up to the present day – and it goes to press just as the discussions about the Challenger 2's future form begin to enter the stage where choices and decisions are made. It is two amateur historians' estimation of a complex story that stretches back to the late 1970s and it will continue to unfold over an exceptionally long service life. Where documentary evidence exists we have noted sources but all gun performance data is approximate and unconfirmed. Our work has benefitted from the kind help of many who have known British tank and component design well and from soldiers who have participated in its design. While we have discussed many parameters of the Challenger 2's technical development, we have also deliberately avoided discussion of factors that might affect the security of crews. We ask readers to understand this very appropriate constraint on our study of a current weapon system.

*(U.S. Army photo by B. Fletcher)*

# Introduction

British tank design practice in the years after 1945 followed the lessons of six years of mortal combat against an enemy who took tank design and tank warfare to a high art form. Operational experience in the last years of the war had deeply influenced post-war tank design and, since the A41 Centurion of 1945, British tank design philosophy prioritized firepower, protection and mobility (in that order). The Chieftain was Britain's first Main Battle Tank (MBT) and, continuing a strong balance of firepower and protection, it became the first NATO MBT to mount a 120mm gun; it also incorporated heavier armour than contemporary European tank designs. The Chieftain's development began in 1951, prototypes were built from 1959 and the first production vehicles entered service in 1967. This 16-year gestation period was typical for the time but it exceeded its planned service life of approximately 20 years. There was no smooth transition to a newer and much more powerful vehicle to replace the Chieftain after 1985. The vehicle we know today as the Challenger 2 was actually developed as the Chieftain's replacement, and it entered service over 15 years after the Chieftain's designed replacement date. The cancellation of one major weapons programme and the chequered career of another were major factors that extended the Chieftain's career into the early 1990s – along with geopolitical factors that nobody could have predicted a decade earlier.[1]

Challenger 2s of the 1st RTR demonstration squadron crossing an AVLB bridge during Exercise Sabre's Thrust in 2000. (*Tim Neate*)

A Challenger 2 at the Bahna Land Forces Day in the Czech Republic in June 2006. (*Pierre Delattre*)

*His Royal Highness the Prince of Wales*, a Challenger 2 Main Battle Tank from the Royal Dragoon Guards taking part in an exercise on Catterick Training Area in North Yorkshire in 2009. (*Published by the Ministry of Defence © Crown Copyright 2009. Reproduced under Open Government Licence*)

With the turret traversed to the 9 o'clock position we can appreciate the size of the Challenger 2's turret bustle. This vehicle belonged to the King's Royal Hussars and was photographed at Zulu crossing on Salisbury Plain in 2008. (*Tim Neate*)

A Challenger 2 fitted with full Theatre Entry Standard (TES) armour kits and mobile camouflage system photographed in late 2016. (*Published by the Ministry of Defence © Crown Copyright 2016. Reproduced under Open Government Licence*)

A 2nd Royal Tank Regiment Challenger 2 photographed at the 2015 Land Combat Power Demonstration. The turret is quite narrow in profile and presents a small and heavily armoured target. (*Tim Neate*)

A Challenger 2 of the King's Royal Hussars photographed at Zulu Crossing, Salisbury Plain Training Area in 2008. (*Tim Neate*)

A Challenger 2 of the Queen's Royal Lancers photographed broken down during an exercise in 2004 after the regiment returned from Operation Telic. The loader's GPMG has been removed from its mounting and the turret roof is cluttered with the crew's gear. (*Tim Neate*)

The rear of the Challenger 2's hull was designed to carry 2 auxiliary fuel drums strapped into brackets fixed to the rear hull plate. These allowed an extra 400 litres of fuel to be carried for approach marches. (*M.P. Robinson Collection*)

The original pattern of Challenger 2 road wheel was very similar to the pattern employed with the Challenger 1. (*Pierre Delattre*)

An improved lighter pattern of road wheels was introduced in the early years of the new century. The Challenger 2's final drives, idler wheels, drive sprockets and tracks have changed little since entering service in 1998. (*Pierre Delattre*)

The muzzle of the L30A1 120mm rifled gun, a weapon of impressive performance that has been used successfully in combat. The top of the muzzle incorporates the muzzle reference system mounting, which allows the gunner to test his sight to gun alignment at any time. (*Published by the Ministry of Defence © Crown Copyright 2016. Reproduced under Open Government Licence*)

With one of its crew huddled on the turret roof, this Challenger 2 of the 1st Royal Tank Regiment was photographed during Exercise Brave Guardian in 1999. By the mud on the vehicle and the attitude of the crewman, we can imagine that the weather was cold and wet at the time the picture was taken. (*Tim Neate*)

Fitted with fire simulator equipment and accompanied by FV432 Mk.2s from Warminster, six Challenger 2s of the 1st Royal Tank Regiment's A Squadron were seen here during Exercise Saber's Thrust in 2000. (*Tim Neate*)

# Chapter One

# Chieftain, MBT80 and the *Shir*

There were two schools of thought in British main battle tank design philosophy in the 1970s. One was the continued evolution of the existing Chieftain design; the other was the development of a completely new MBT design. The second of these trends, which encompassed the Future Main Battle Tank (FMBT) and Main Battle Tank 80 (MBT80) programmes, became the single greatest barrier to replacing the Chieftain in a timely manner. Both developmental trends were pursued in the 1970s and resulted in considerable investment from the government and from private companies in the defence sector. Component, armour and weapons development for official and non-official projects resulted in a wide range of technically brilliant solutions but also to a dilution of effort (aggravated by political meddling at high levels). Britain shared many of its innovations with allied powers in good faith but ultimately the continued development of the Chieftain design was the path forced on the army after the two new MBT programmes failed.

The FMBT programme of the early 1970s was premised on co-production with West Germany and was discarded in 1976 because British and West German design priorities and doctrine were incompatible. The other tank producing nations in NATO had their own conscript armies and tank producing industries and no possible partner had much interest in adopting British design practices. The MBT80 project that succeeded the FMBT programme was also deeply tied to the whole notion of production with a partner nation. Discussion of what the MBT80 might have resembled remains conjectural. What is known is that the most advanced technological solutions to the different elements of MBT design would have been included in a vehicle weighing over 65 tons.[2]

The MBT80 remained a loosely defined paper project throughout its existence, despite Britain having plenty of innovative technology available to create world beating tank designs in the late 1970s. Proposed features for the MBT80 incorporated a number of proven and unproven technical solutions; the more unusual proposals ranged from mounting the main armament in a turret with offset gun trunnions to the possible use of an aluminium hull rear section to minimize weight. It is unlikely that such unconventional features would have been included in a production vehicle but many avenues were expected to be explored and innovative ideas to be tested. A production MBT80 would have carried an advanced 120mm rifled gun developed from the guns tested for the United States Army's MBT main armament programme. It would have been frontally protected with Chobham armour with detachable

Chobham armour side arrays. The engine had not yet been finalized when it was cancelled but might well have been a 16-cylinder version of the Rolls Royce Condor diesel giving it unprecedented power.[3]

For all of the innovations included on the wish list of technical features for a new MBT, the most important extended to the fire control system, which was expected to include the most advanced panoramic thermal vision systems yet devised by British companies like Barr and Stroud. Despite the extensive paper studies and the two test rigs built to research the design parameter of the MBT80, Britain lacked the funds and the political will to develop a definitive Chieftain replacement.[4] There were additional factors that made the MBT80 programme extremely vulnerable to cost overruns and to delays in finalizing the specification. The MBT80 programme became a political morass by the late 1970s given government pressure from the highest levels to find an allied production partner nation to share the development costs. The Americans, West Germans and French had their own programmes and expressed little interest in joining the British despite entreaties at the highest levels. Nothing resulted and by the time it was cancelled a definitive prototype had not even been constructed. Given the economic and political crises that marked the late 1970s it is easy to see that the Chieftain replacement was but one of many defence priorities the government had to review.

The second (and far more practical) trend of tank design being undertaken in the UK in the 1970s was being pursued by Royal Ordnance and Vickers for export clients. Vickers marketed their 37 ton Vickers Medium Tank, which was produced under licence in India, to African armies and Kuwait; Royal Ordnance marketed the Chieftain and developed that basic project into a much improved design for the Imperial Iranian Army. The ultimate version proposed to the Iranians transformed the basic Chieftain design into the *Shir 2* by the late 1970s. The definitive *Shir 2* was driven by a 1200 horsepower Rolls-Royce diesel and featured a hydro-pneumatic suspension. It employed a welded armour steel turret and hull with the turret front and glacis protected with revolutionary Chobham composite armour. The *Shir 2*'s composite armour layout closely followed that proposed for the experimental FV4211 'Aluminium Chieftain'. The design represented great advances in protection and battlefield mobility over the British Army's own Chieftains but retained the 120mm L11A5 gun and the IFCS fire control system that was standard on the Chieftain Mk.5. Other British firms had perfected advanced night vision equipment, more modern fire control systems and advanced sighting equipment. The British thus had every ingredient for world class battle tank design available from their domestic industries.[5]

The decision to stop the MBT80 programme and instead to procure a modified *Shir 2* was controversial and had political overtones but it was the express wish of the Royal Armoured Corps – the "user" – as promoted vehemently and successfully by the relevant OR Branch in the MOD. The reasons were obvious; it was expected greatly to exceed the original cost estimates and require an unacceptably long time before completing its development phases and entering service. The requirement for a Chieftain replacement was urgent, partly due to NATO's exaggerated impression of the Soviet T64 and T72's capabilities. A second factor was the change of regime in Iran and the cancellation of the Iranian *Shir* orders. The Iranian affair left Royal Ordnance on the precipice of fiscal disaster and the British government with an opportunity to exit the MBT80 programme; the *Shir 2* was thus bought practically 'off the shelf'. While the *Shir 2* design was short of the General Staff Requirement intended for the

MBT80 its cost was about 40 per cent less per unit and the sound design included a high proportion of tested components. It was expected to be available for issue within 5 years and, with a list of modifications, it met army approval; eventually 400 of these vehicles were ordered to equip nearly half of the MBT regiments in the British Army of the Rhine. With new thermal vision devices and other modifications to make them suitable for British Army service, the *Shir 2* design became the British Army's FV4030/4. Christened 'Challenger', the new vehicle was not intended to replace the Chieftain, but rather to supplement the existing MBT fleet.[6]

In the 1980s, therefore, the Royal Armoured Corps was condemned to operating two types of MBT. The Chieftain was fast nearing the end of its life; even after its upgrade in the early 1980s with Stillbrew armour and the Thermal Observation and Gunnery System (TOGS) introduced with the Challenger it was still basically a 1950s design. Feasibility studies had been conducted since 1969 to up-gun and up-armour the Chieftain. For the most part upgrades and modifications had focused on increasing the Chieftain's armour protection, its ammunition stowage and its power output. The Chieftain's notoriously unreliable engine became the focus of most of the available funding for the MBT's modestly successful improvement in the 1970s. By the 1980s the suggested improvements to the Chieftain offered limited returns as advances in armour, guns and in fire controls could be delivered on a newly built MBT.[7]

A Chieftain Mk.6 or Mk.9 of the 15th/19th The King's Royal Hussars at Bovington. This vehicle was built in the late 1960s as a Mk.2 and like most Chieftains, it was upgraded repeatedly. There were nearly 1000 Chieftains in British service in the early 1980s, and the last were withdrawn in 1994-95. The Chieftain was theoretically already past its scheduled replacement date in 1986 when this photo was taken. (*Tim Neate*)

A Chieftain Mk.11 of the Royal Scots Dragoon Guards in the 1980s. The Mk.11 was updated with the TOGS system introduced with the Challenger 1 and it carried Stillbrew appliqué armour. These measures provided a more capable MBT to the armoured regiments equipped with the Chieftain in BAOR but only 200 vehicles were upgraded to Mk.11. The Chieftain Mk.11 still suffered from a lower power to weight ratio than many of its contemporaries. (*Tim Neate*)

The FV4211 Aluminium Chieftain was the vehicle from which the Challenger 1's armour layout developed. The FV4211 was the first MBT designed to mount Chobham armour. (*Keith Paget*)

This photo of the glacis arrangement of the FV4211 clearly shows the glacis armour arrangement inherited by the Shir 2, Challenger 1 and Challenger 2. (*Keith Paget*)

Seen here as a derelict, this is the ATR2 test rig, numbered 99 SP 27, the only surviving prototype vehicle from the MBT80 programme. Two test rigs were built ahead of the proposed construction of MBT80 prototypes. The details of their brief careers as test vehicles are unknown. (*Keith Paget*)

The ATR2 is a vehicle shrouded in mystery because no documentation has yet emerged to confirm its exact place in the MBT80's development. We can see a general similarity with the *Shir 2* hull layout, but the gun mantlet features a wider trunnion arrangement than a Chieftain turret. (*Keith Paget*)

A Challenger 1 prototype photographed on a Royal Ordnance display stand in July 1982. (*Tim Neate*)

Seen here in 1983, a Challenger 1 prototype puts on a display at Bovington. The production Challenger 1 featured a TOGS barbette on the front right of the turret next to the gunner's primary sight. (*Tim Neate*)

This 1984 photo shows a very early Challenger 1 (probably a Mk.1, the designation given to vehicles issued prior to the TOGS systems being fitted). The first unit to receive these tanks was The Royal Hussars. (*Tim Neate*)

The Chieftain 900: a demonstrator developed in the mid-1980s that effectively proved the feasibility of rebuilding the Chieftain with the Marconi Centaur FCS, Chobham armour protection and a 900 HP powerpack. It was never adopted by the British Army, but it also showed that the Challenger 1 might have been adapted to a better fire control system. (*Tim Neate*)

## Chapter Two

# Challenger 1

The *Shir 2* was a fine tank by any measure in 1980. It was in essence an improved Chieftain turret design mounted on a larger hull with a new powertrain and suspension. It had a proven L11A5 120mm rifled gun; its Chobham armour protection would have proven very difficult to knock out frontally. It was a far more mobile vehicle than the Chieftain and looked like a modern wonder of military technology but not all features were new. The simple fact, however, was that the off-the-shelf *Shir 2* was the most advanced British MBT design available in 1980. In hindsight the decision to buy the FV4030/4 (referred to hereafter as the Challenger 1) was correct. The *Shir 2* design was based on sound principles and could enter service within a reasonable timespan with a turret incorporating proven subsystems.[8]

The Challenger 1 was not without faults. The threat of privatization meant that Royal Ordnance was desperate to sell it abroad, and the publicity campaign that followed its entry into service touted it as 'the best tank in the world'. For all its new features, it also incorporated old technology and older style electronics, particularly in its turret systems. It above all lacked the new generation panoramic stabilized thermal sights which would have been the heart of the MBT80's fire control system and which already featured in the Leopard 2 and M1 Abrams ordered by West Germany and the United States. A smaller problem to the British upon adoption of a second MBT alongside the Chieftain was that it required a second complete supply chain of mechanical and suspension spares. None of the Chieftain's parts were compatible with the Challenger's, so the RAC's logistics were effectively doubled – a costly proposition that resulted in an inadequate stock of spares being provided both in the UK and for the British Army of the Rhine. In addition to suffering the inevitable teething troubles of a new weapon system this lack of spare parts was a source of embarrassment for the new Challenger regiments.[9]

The government was well aware of the Challenger 1's availability problems and various options were considered to address the issue as economically as possible. One solution was to buy more Challengers to standardize the British MBT fleet and enable the Chieftain to be retired. This would have addressed the reliability issues but at substantial cost. Other suggestions included the unpalatable option of buying a foreign design to replace the Chieftain. Senior RAC officers rejected the foreign MBT option but the idea long persisted amongst politicians and even within the army where the poor reliability of both Chieftain and Challenger was a source of frustration. Recalling that the notion of co-producing an advanced main battle tank with another NATO nation had been entertained throughout the 1970s, the idea of buying a tank whose development and teething problems had already been worked out was a logical further step in the same direction. The Leopard 2 had greatly impressed many British officers due to its reliability and excellent overall design. The Royal Air Force had long employed foreign designs, and some politicians considered that the same rationale could logically be applied to the army's tank procurement.[10]

The notion of national pride in an army that had fielded the first tank was not the only reason why the British Army could not enthusiastically embrace a foreign built design. Available foreign MBT designs were not ideally suited to British needs. In the case of the otherwise excellent Leopard 2A4, the armour protection was well below the minimum British requirement. The American M1A1 was protected with Chobham armour but rejected because its thirsty gas turbine power-plant did not suit British requirements. Both of these foreign types offered NATO interoperability, and mounted the smooth bore Rheinmetall L44 120mm gun, but the RAC stood firm on the need for a rifled main armament.[11]

The RAC's main objection to the Rheinmetall gun was its limited range of ammunition. The smoothbore L44 120mm gun was principally developed as an anti-armour weapon firing kinetic rounds on a very flat trajectory with excellent armour penetration. At the time of its introduction, the 120mm smoothbore's armour piercing round was equivalent to the performance available from the British 120mm rifled APFSDS rounds developed for the L11A5 series guns, but the German weapon was not provided with a satisfactory high explosive round. The performance of early 120mm L44 smoothbore high explosive rounds was considered by the British to be very poor compared to the existing 120mm High Explosive Squash Head (or HESH) ammunition developed for the L11 gun – a type of round long favoured by the British. HESH was useful against a wide range of targets and combined good blast characteristics with excellent anti-armour performance. The use of a rifled gun was a parameter that the RAC refused to abandon even in the name of NATO standardization.[12]

# Chapter Three

# The Vickers Mk.7

The official Ministry of Defence procurement programme for a new MBT to replace the Chieftain began anew in November 1986, and Vickers's newest MBT design showed considerable promise as a possible contender. Vickers had not been affected by the same problems that bedevilled Royal Ordnance in the early 1980s and was doing well in the export market. Vickers was not wedded to the notion of 50-plus ton tank designs themselves, but they had a 120mm-armed MBT turret under development since the early 1970s; this was first shown on the aluminium hulled Vickers Valiant and developed by 1984 into a vehicle marketed as the Vickers Mk.7 MBT. The 54 ton Vickers Mk.7 was a private venture that Vickers had pursued with the West German Krauss-Maffei company of Kiel (who supplied the Leopard 2 hull complete). West Germany had avoided selling arms in the Middle East up to that point and the Mk.7 concept capitalized on using a Krauss-Maffei hull with a British turret. Vickers attempted to sell the Mk.7 MBT to middle-eastern clients in June 1985 without success but the design caught the army's attention. The Mk.7's turret offered a range of options for sights and for the main armament (including the 120mm L11A5, the 120mm Rheinmetall L44 and the GIAT 120mm G1, which fired a round compatible with the West German gun). The Vickers turret came from an entirely new design concept that was a complete departure from the FVRDE/Royal Ordnance Chieftain-*Shir 2* evolutionary path.[13]

    The Mk.7 turret was built around a hunter-killer fire control system that the Chieftain/*Shir 2* turret lineage lacked. This type of fire control system features a panoramic (traversable) stabilized sight for the commander and a stabilized sight with a thermal imager for the gunner. The design of the turret and its fire controls had been refined since the Vickers Valiant universal turret of the late 1970s. The Mk.7 turret already incorporated proven subsystems for the commander and gunner (which included the commander's sight from SFIM in France and the Philips UA9090 thermal imaging system from the Netherlands). The RAC's Armoured Development and Trials Unit (ADTU) provision of crews for the Mk.7's trials in the Middle East in 1985 increased the army's awareness of its capabilities. By the time Vickers took the Mk.7 for trials in Egypt the tank was already well-regarded for its innovative and amazingly fast fire control system.[14]

    The Vickers Mk.7's integrated Marconi Centaur 1 digital fire control and gun control system included a modern ballistic computer. It departed from the archaic electro-mechanical gun stabilisation method used since the Centurion (with servo motors stabilising the gun itself) to the new concept of the commander's sight mirror itself being gyrostabilised in its line of sight. The turret and gun were slaved to the independent commander's sight or to the thermal sight electronically. Electronics replaced old and unreliable mechanical sight linkages fundamental to the Challenger 1's fire controls.[15]

In November 1986 Vickers first proposed an MBT based on the Vickers Mk.7's turret mated to an upgraded Royal Ordnance Challenger hull as a private venture. The hull would boast a new transmission, an improved suspension and a range of other improvements. This made the most of two proven designs and coincided with the purchase by Vickers of Royal Ordnance's tank manufacturing plants. The company ran a presentation to the Ministry of Defence elaborating the project as a Chieftain replacement on 30 March 1987.[16]

The Vickers Mk.7 MBT was the starting point for the Challenger 2's turret. The Mk.7 was never adopted by any foreign armies despite demonstrations in Egypt and elsewhere in the Middle East in 1985. The turret design featured the GEC Marconi Centaur hunter-killer fire control system. Using improved components, this fire control system evolved into that of the Challenger 2. (*Vickers Defence*)

Chapter Four

# GSR 4026, CAT 87 and the End of the Cold War

Following the successful Vickers presentation, General Staff Requirement 4026 entitled *Staff Requirement for the Replacement of Chieftain* (hereafter referred to as GSR4026) was issued by the British Army on 30 November 1987. This put the Vickers-sourced MBT project on a formal footing as a contender for the Chieftain's replacement and, due to events, it gained urgency very quickly. The traumatic episode that drove GSR4026 hardest was the Challenger 1's poor performance at the NATO CAT87 gunnery contest.[17]

The Canadian Army Trophy competition was the most prestigious gunnery contest in NATO. The ethos of the event was that each participating army's teams should be selected from serving armoured regiments and that current equipment would be used to ensure fairness. British teams mounted on the Centurion had won the CAT Trophy in 1965 (the Royal Scots Greys) and again in 1966 (the 13th/18th Hussars). The Chieftain had done reasonably well in the 1970s, coming first in 1970 (the 16th/5th Lancers) and second in 1973 (the Queen's Royal Irish Hussars).[18]

By the late 1970s the old values held in the CAT competitions were long gone. In 1979, the *Bundeswehr* team trained for twelve months prior to the CAT competition, and their team's Leopard 1A4 tanks were looked after by Krauss-Maffei. Other competing NATO army teams began to use vehicles maintained only by their own fitters. The trend became for manufacturers to become increasingly involved. The later CAT competitions were conducted with serious political overtones and with very heavy national gamesmanship. By 1987 the Canadian Army Trophy competition had become a platform for selling weapon systems, and no tank producing nation wanted to miss the chance to sell its wares.[19]

Before the June 1987 contest the Ministry of Defence and the Royal Armoured Corps convinced themselves that the Challenger could win. Two British regimental teams trained for CAT87, the Royal Hussars and the 2nd Royal Tank Regiment. Eventually the Royal Hussars were picked to compete and their tanks received what can only be described as minor modifications to their fire control systems to give them an edge on the newer and more advanced systems used on the West German Leopard 2A4 and the American M1A1. The Royal Hussars crews were well trained and had full confidence in their tank but events did not favour the British. CAT87 was all about quick gunnery and target acquisition: a showcase for advanced fire controls. The other factors favouring the Challenger 1 as a battle tank – excellent armour, a powerful gun, steadiness as a gun platform and battlefield mobility – all counted for nothing in the competition. The failure of Challenger 1 to shine at CAT87 proved a disaster for British prestige as a tank producing nation and the design was totally discredited.[20]

The British government had forgotten that it had bought the Challenger off-the-shelf in a time of crisis. The army had seemingly also forgotten the limitations of the Challenger's design. The tank, with its 1970s-era fire control system, entered a highly publicized competition against others with newer technology, fire control computers and panoramic stabilized sights. The Challenger was relatively slow to acquire targets and had a low hit rate during the competition. One can only imagine how the stress of trying to speed the rate of fire to compensate for the slower fire control system took its toll – and how hard the Royal Hussars crews had struggled with the Chieftain-inspired layout of the Challenger turret. Blame was heaped on the L11A5 gun, on the IFCS system and on the turret ergonomics. The Challenger's old fashioned features suddenly engendered the gravest doubts compared to the technical excellence of the turret layout and fire controls of the Leopard 2A4 and M1A1. The RAC had seemingly harvested the bitter fruits of the MBT80's cancellation. A solution to the mess was demanded immediately at the highest political levels and the media lost no opportunity to attack the Ministry of Defence and the army. While a solution was sought the RAC simply had to make the best of what they had.[21]

Recriminations flew around the corridors of power following CAT87 adding credibility to those who had championed buying a foreign tank to replace the Chieftain. Neither the Leopard 2A4 nor the M1A1 design fulfilled the British Army's General Staff Requirement although the Leopard's reputation for reliability bought it support within the army. The criteria for the staff requirement could only be met by the foreign designs if the M1A1 and Leopard 2A4 were drastically modified, and because the Vickers proposal was still only a paper design it was met in some quarters with scepticism (including some Ministry of Defence circles) because a private company was developing a tank with minimal MoD direction. Vickers refined their MBT proposal and presented it to the Ministry of Defence in February 1988.[22]

Prime Minister Margaret Thatcher realised that a foreign purchase could be political suicide (especially in terms of lost jobs) and that it would sound the death knell for one of Great Britain's strategic industries. She vetoed any such suggestion and demanded that the MoD look again at the options. Vickers had acquired Royal Ordnance's tank building facilities immediately before the Challenger 1's design limitations were highlighted at the Canadian Army Trophy competition. As a result of Vickers's significant experience in developing solutions for foreign armies, and because it was now the only company capable of producing an MBT in Britain, the Ministry of Defence had a political stake in its continued success.[23]

To aid Vickers in the development of their proposed MBT, a £90 million contract was issued in December 1988 to cover the demonstration and proof of principle development phase. Later a portion of the development costs incurred by Vickers during the Mk.7's development were also reimbursed. The government's requirements in a new MBT specification included vital performance criteria that had to be met by the Vickers design. These included reliability, ease of maintenance and ergonomics. Vickers' own design team had already placed great emphasis on ergonomics and systems reliability in the Mk.7 turret. To achieve this Vickers was willing to source subsystems from foreign suppliers if it guaranteed the tank a more reliable option than could be obtained from a British contractor. As a result, the French SFIM commander's sights first seen on the Mk.7 were retained, and the gunner's sights were produced as a SFIM-Barr and Stroud cooperative venture. Vickers knew that its ability to gather government and army support for its design came down to reliability and demanded the same from its suppliers. Following this theme the turret systems that evolved from the Mk.7 turret into the eventual Challenger 2 production tank saw many component substitutions as more reliable options became available.[24]

The M1 Abrams was the victor of CAT 87, equipping both the first and third place teams (including the vehicle seen here from 3-64 Armor, A Company, 1st Platoon). The design concepts behind the Challenger 1 and the M1 series were quite different (despite the heavy armour emphasized in both designs). The later M1A2 was a strong contender for the Chieftain replacement, although its gas turbine powerpack was a feature that the British Army had already studied for the MBT80 and did not favour. The fire control computer later adopted in the Challenger 2 design was similar to that of the American tank, selected for its ease of use and reliability. (*Ron Mihalko*)

15

The Royal Hussars CAT 87 patch. The Canadian Army Trophy gunnery contest of 1987 was a humiliating experience for the Royal Armoured Corps, and one which generated a great deal of doubt in the British Army and government about the British defence industry's ability to develop a world beating main battle tank design. Vickers's choice of numerous foreign subsystems in the Challenger 2's fire control system was one consequence. The Vickers's design team placed the highest emphasis possible on reliability in every system of the Challenger 2, adopting proven components from abroad where necessary to ensure that the MBT had a reliable, user-friendly fire control system. (*Ron Mihalko*)

The turret layout inherited by the Challenger 1 from the Chieftain is plain to see. The Challenger 2's most important improvements over this design were in modern electronic fire controls and in the provision of a traversable (or panoramic) commander's sight employed to find targets for the gunner quickly and efficiently. (*Lawrence Skuse*)

A Soviet T64 being examined by British soldiers after the end of the Cold War. The end of the Cold War had a significant impact on how the Chieftain was replaced and also impacted Vickers ability to sell the Challenger 2. (*Lawrence Skuse*)

The Leopard 2 A4: a splendid piece of Cold War engineering that entered service in 1979 and attracted a significant amount of support in the RAC as the Chieftain's potential replacement. (*Jurgen Scholz*)

# Chapter Five

# The 120mm L30A1 gun

In the mid-1980s the CHARM (Chieftain/Challenger Armament) project was undertaken to permit a wholesale main armament upgrade of the existing British Army Chieftain park as well as the entire Challenger 1 fleet. The L11A5 gun carried by the later and upgraded Chieftains and the Challenger 1 was a potent weapon but an even more powerful 120mm rifled gun that could be fitted to the existing tank fleet was intended, to keep the Royal Armoured Corps abreast of gun development in the Warsaw Pact. The L11A5 gun already in service was to be provided with a new armour piercing fin stabilized discarding sabot tracer (APFSDS-T) round with a depleted uranium penetrator (which entered service as the L26A1 in 1990). Vickers successfully bid as prime contractor for the CHARM programme, which also included co-development with Royal Ordnance (ROF Nottingham) of a new 120mm rifled high pressure gun to fire a yet more effective kinetic round also with a uranium penetrator. The CHARM gun emerged in 1989 as the L30A1 120mm rifled gun. The L30A1 is a 55 calibre length, chromed bore gun with a split breech block and reputedly with a barrel life of approximately 450 rounds. The ammunition for the L30A1 continued the L11 120mm gun's use of separate charge ammunition. Each round was loaded in 3 pieces, first the projectile, followed by a combustible bag charge and finally a vent tube to ignite the charge. This allowed the Challenger 2 to fire most of the older L11 rounds as well as the new CHARM ammunition. While the multiple part ammunition system is more complicated than that used on the 120mm smoothbore guns, it meant that all that came back into the turret on recoil was the small calibre vent tube ejected from the breech. The 120mm smoothbore guns have a stub case at the base of the charge and the charge and projectile are loaded as one, similar to an old brass cased round, so that on firing the case is burnt and the stub case sealing the breech during firing is ejected.[25]

Like the L11, the L30A1 has obturators fitted into the breech to provide the seal although the breech was a new design. The mechanism employed on the L30A1 is a split vertical sliding-block breech. One vertically sliding block holds the obturator ring and is locked for firing by a second block. When the second block falls, the first is released to open the breech. The decision to develop the L30A1 was taken after other options were examined. A 140mm rifled gun was also designed during the first half of the 1980s, firing 2 part ammunition. This gun was built and test fired from a test mounting but is not believed to ever have been fitted into a tank turret. A small number of Chieftain test vehicles were used by the MVEE to establish the gun control characteristics of the larger weapon on a tank of the 50-60 ton class. The ergonomics of handling the large 140mm rounds and the reduction in ammunition capacity brought back memories of the British 183mm gun projects of the 1950s. While technically viable, even in the 1980s, the use of 140mm weapons presented serious tactical problems. Further development of a rifled 140mm gun was dismissed in favor of enhancing the 120mm rifled gun.[26]

The critical date for Vickers came on 30 September 1988 when representatives of the company went before a parliamentary committee to prove their new tank would be a suitable contender for the RAC. A factor in favour of their bid to secure the contract for the Chieftain replacement was their involvement in the CHARM project. It was already intended to procure the L30A1 gun for the CHARM programme and the government intended to keep an upgraded Challenger 1 in service after the Chieftain was replaced. The new Vickers tank would also mount the L30A1 gun. Development proceeded over the next two years at Vickers and the government decided on a comparative MBT trial to decide the Chieftain's successor.[27]

# Chapter Six

# The New Vickers MBT

The turret configuration that was selected for the new Vickers tank followed closely that proven on the Vickers Mk.7 MBT. A similar layout had been pursued in the turret developed by Vickers in 1985 for the EE T1 Osorio for the Brazilian Engesa company. Vickers favoured a wide external mantlet which allowed maximum trunnion spacing for a more stable gun mounting than was possible for the Chieftain and Challenger gun mountings. The wide set trunnions enabled the mantlet mounting points to be far more substantial and helped mitigate the shock absorbed by the trunnions when firing high pressure ammunition. A long study of this problem was conducted by Vickers on the issue of trunnion shock that predated both the Mk.4 and Mk.7 turrets.

Ammunition stowage for the new tank followed the practice established on the Challenger 1 Mk.3. APFSDS rounds were stored on the turret walls and on the turret sill. A ready rack was provided in the turret bustle. Explosive ammunition and charges were stored below the turret ring; charges were kept in armoured bins located around the rear of the fighting compartment, on each side of the driver, and in the hull corners. Approximately 50 rounds of main armament ammunition were stowed although the ammunition selection depended on the tactical mission undertaken. Besides L30 HESH and L23 APFSDS rounds, the L30A1 also fired new CHARM APFSDS depleted uranium rounds (L26A1 CHARM 1 and L27A1 CHARM 3), and smoke rounds were also available. The Royal Armoured Corps had favoured the High Explosive Squash Head or HESH round ever since the L7 105mm rifled gun had been introduced in 1959 as its multipurpose explosive round, and it was on this point that the whole debate between the 120mm rifled and smoothbore weapons hung.[28]

The co-axial weapon, like that specified for the Valiant and the Mk.7, was the McDonell-Douglas Helicopter (later Hughes) L94 7.62mm Chain Gun, a complete departure from two generations of British MBTs armed with coaxial machineguns derived from infantry weapons. The ammunition and charges were stored in armoured bins below the turret ring level. The turret roof was as clean and unobstructed as possible to permit use of the panoramic sight mounted on the commander's cupola, functioning something like the periscope of a submarine to identify targets which could then be passed to the waiting gunner. The hunter-killer fire control system offered a tremendous advantage in both sight range and speed of engagement to the turret crew.

From 1987 until 1990 Vickers pursued the development of the Chieftain replacement by combining the Mk.7 turret with the Challenger hull inherited from its purchase of Royal Ordnance. This process resulted in many changes to both of these major assemblies. In the case of the hull an improved Dunlop hydropneumatic suspension and an entirely new David Brown TN54 epicyclic transmission were fitted. These major components were already being perfected for the Challenger Armoured Recovery and Repair Vehicle (CRARRV) which Vickers was contracted to build for the British Army in 1985.[29] The Mk.7 turret received substantial modifications to the fire control system and to its armour protection. The army also had a requirement to permit more rapid gun removal than was possible on the Chieftain and Challenger 1. On the Challenger 2's gun mounting the breech ring can be supported, the front barrel support bearing is unbolted and the barrel comes out of the front. In the original Mk.4 Valiant and in the Vickers Mk.7, the gunner's primary sight was the simple Nanoquest L30 telescopic sight mounted in the mantlet and incorporating a laser rangefinder, with a secondary periscopic sight on the turret roof immediately above the gunner's position. The RAC's requirements included that the periscopic sight be the gunner's primary sight, which Vickers resolved by contracting the assembly to Barr and Stroud, who in turn subcontracted the sight head unit to SFIM. The resulting sight was simply designated the 'Gunner's Primary Sight' or GPS, an assembly particular to the Challenger 2. The telescopic sight featured on the Mk.7 was not originally carried forward to the Challenger 2 turret design, but was added on recommendation of the Armoured Demonstration and Trials Unit crews who had put the Vickers Mk.7 through its paces. The L30 telescopic sight, in its simplest form, was retained on the production vehicles as a secondary sight fitted with an armoured cover flap in the mantlet's right side.[30]

A second major change adopted during this time was the replacement of several major electronic components in the fire control system demonstrated on the Mk.7 with proven assemblies that had already been extensively tested elsewhere. One element that was replaced was the Marconi Centaur 1's fire control computer; the army preferred the Canadian manufactured CDC fire control computer which had already proven itself on the M1 Abrams. The SFIM VS580-10 commander's panoramic sight first shown on the Vickers MBT Mk.4 Valiant (and brilliantly demonstrated as the most important part of the Mk.7's optics during the type's trials for the Egyptian Army in 1985) was retained for the new tank. The Mk.7's commander's cupola was replaced by a version with 8 periscopes but retained its layout with a button under each permitting traverse of the turret to bring the gun to bear on the commander's line of sight. The armour specified for the new Vickers tank was an improved version of the Chobham armour fitted to the Challenger 1 known by its code name Dorchester.[31]

Wearing Challenger 1 style bazooka plates and tracks, prototype 06 SP 93 closely resembles the original Vickers concept of 1987: a new turret on the Challenger 1 hull, although the prototypes had the same range of hull improvements intended for the production vehicles. Two of the prototypes were used for mechanical reliability tests at ADTU Bovington, one (V9) was retained as a demonstrator for foreign sales, and became the Challenger 2E. Today some of the prototypes survive as gate guardians. (*Vickers Defence*)

In this view can be seen the Commander primary sight fitted to the turret of prototype V5, as well as the gunner's primary sight under its protective hood and the gunners unity vision block, which was fitted after user feedback to give him a little more view of the outside world without total reliance on the gunnery sights. (*D. Lunn*)

The two large cylindrical objects visible in this shot are the electric motors that power the cupola periscope wipers on prototype V5, a vehicle we can suspect was at some point user tested by the Royal Scots Dragoon Guards, and which still carried the Scottish saltire on the TOGS mounting sides at the time these photos were taken. (D. Lunn)

This photo of prototype V5's turret roof taken inside a tank hangar gives a good view of the early loader's machine-gun mounting, the first ever fitted to a loader's hatch on a British MBT. The need for an uninterrupted field of view for the commander's panoramic sight was the main reason for the adoption of this weapon on the loader's position. The original loader's machine-gun mounting was a steel tube affair of dubious appearance, rather akin to the sort of mounting seen on the observer's cockpit of a Royal Flying Corps biplane! The second type, which was adopted only after production began, was much simpler in design and ultimately far more suitable to the task. (D. Lunn)

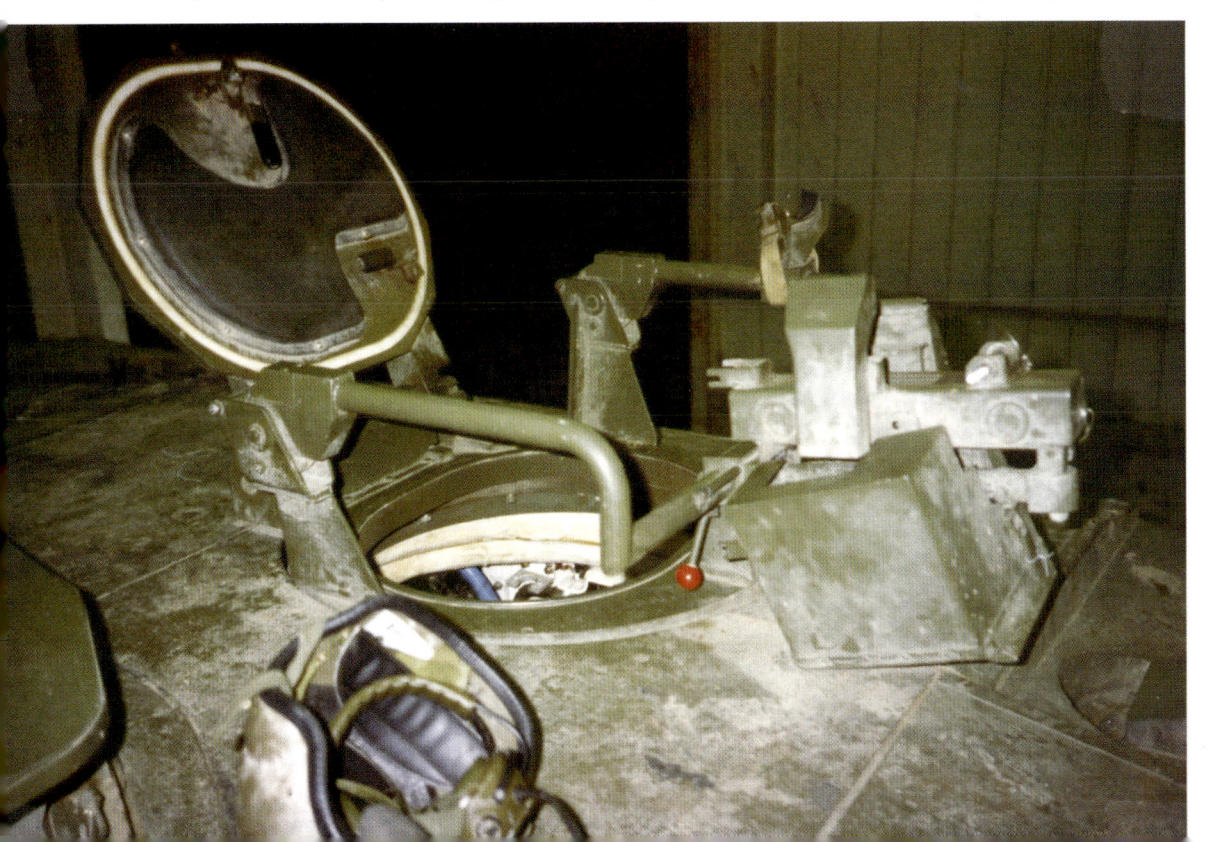

# Chapter Seven

# Options for Change, Operation Granby and the Selection of the Vickers Challenger 2

Larger events formed the backdrop to the final development of the Challenger 2 as Vickers began construction of prototypes. The Warsaw Pact unravelled in 1989 and the Soviet Union effectively ceased to exist as its republics broke away in early 1990. NATO reacted by watching, waiting, and very quickly reducing military spending in turn. The British government conducted the Strategic Defence Review in late July 1990, which changed government defence policy to reflect the change in strategic outlook since the end of the Cold War. The policy document prepared by Secretary of State for Defence Tom King, and known to history as 'Options for Change', recognized that, with the disappearance of the Warsaw Pact, a conventional war in Europe was most unlikely for the foreseeable future. This presaged a massive shift in defence policy permitting an 18 per cent reduction of all British armed forces.

The implications for the Royal Armoured Corps were significant and amalgamations, resulting in regiment numbers being reduced from nineteen to eleven, were expected to be complete by 1993. As a consequence the number of MBTs that the British government intended to purchase to replace the Chieftain was reduced almost immediately, a factor that would naturally impact the decision of which MBT to buy. The army, wary of the recently fallen Iron Curtain, had before summer 1990 intended to acquire approximately six hundred new MBTs, to serve alongside the Challenger 1. The Chieftain had always been intended for this role but this was dropped with the reduction of the size of the RAC as directed in 'Options for Change'. The new tank would now be a wholesale replacement for both the Chieftain and the Challenger 1. The Germans and the Americans rightly saw this as a factor that could force the British to discount the option of buying a foreign tank. The competition was open to Vickers, the United States, Germany and France – the only other tank producing nations able to meet the requirements of the GSR. The decision to amalgamate so many units caused much discussion within the RAC but the process was to be staged over 3 years, and, as soon as 'Options for Change' was announced, an opportunity arose to test the Challenger 1 in combat.[32]

In August 1990 Kuwait was invaded by Saddam Hussein's Iraqi army, and in late 1990, when the United Kingdom decided to take part in driving out the Iraqis, it became obvious that the

British had to deploy an armoured division as part of the coalition forces. The discredited Challenger 1 was to receive a baptism of fire against what was expected to be a well-armed enemy. Dismal predictions resulted in some British newspapers and a number of stories dragging up the results from CAT87 inevitably followed. As the 1st Armoured Division shipped their vehicles to the Persian Gulf, the battle lines for the competition to find the British Army's next MBT were finalized.[33]

The Gulf War vindicated British tank design philosophies and the performance of Vickers supporting Operation Granby was counted favourably by the government. Vickers brought the CRARRV recovery vehicle into service during the operation; it was a great success in terms of reliability and design. The Challenger 1 redeemed itself when put to the test of battle; it proved to have a high reliability rate when provided with an appropriate scale of spares. It suffered no losses and entered the record book with the longest range tank kill ever recorded. The salvage of its reputation also favoured a Vickers tank by association. The Challenger 2 concept had also benefitted from strong support within the British Army purely on its own merits.[34]

The MBT procurement competition was delayed until July 1991 because the entire British tank industry was expected to provide technical support for Operation Granby. One area of concern was that the other western tank producing nations did not immediately jump to offer their tenders. The idea of competitive trials deciding a main battle tank procurement programme had not been attempted since the 1970s in any major NATO army – both Germany and the USA suspected that the British trial would not be conducted objectively but, after numerous high level meetings, the Germans and Americans were convinced that the competition would be conducted without bias. Competing vehicles were to be assessed and crewed by selected soldiers from the RAC to enable each to be tested versus the army's requirements. Germany's Krauss-Maffei proposed the new Leopard 2A5 with its improved turret (addressing the armour protection issues identified on the Leopard 2A4). The USA's General Dynamics entered an improved version of the excellent M1A1. A latecomer in the competition, and by far the most unconventional, was the French Army's GIAT Leclerc.[35]

The M1 and Leopard 2 were similar in concept to the proposed Challenger 2 in that all had four-man crews and similar layouts for fire control systems. The futuristic GIAT Leclerc was the odd one out but neither its three man crew nor automatic loader were features favoured by the RAC. After the trial the Leclerc was dropped from consideration (due to its automatic loader, its smaller size and its armour protection not meeting the GSR) leaving the Leopard 2A5, M1A2 and Challenger 2. These remaining three all met the staff requirement and each had particular strengths and weaknesses. The British tank had the lowest horsepower per ton of the three but it addressed all of its predecessor's failings and showed great promise for further development. It must be remembered that the competition was an evaluation of each tank versus the GSR rather than against each other.[36]

The final choice was left to the British government, but after Operation Granby, the whole geopolitical outlook had changed. Since the days of Vickers's first proposal for a Chieftain replacement, the Cold War had ended and the prospect of large orders had disappeared. Prime Minister John Major decided in favour of Challenger 2, but the process had a significant caveat: Vickers would be subjected to a very strict set of acceptance conditions if they were to be allowed to put the tank into full production. The official name 'Challenger 2' was adopted in order to deflect MoD critics; the name implied that the new tank was an improved version of the Challenger 1 and thus side-tracked attention from the as yet incomplete deliveries of new Challenger 1s. The formal announcement to the House of Commons took place on 21 June 1992 and the contract for 127 tanks and 13 driver training tanks (DTT) was awarded to Vickers Defence. The Americans were by far the most vocal in their disappointment at not securing the British MBT order and demanded an explanation at the highest levels. For Vickers the government's decision to adopt the Challenger 2 was a resounding success and endorsed the company's persistence and the resourcefulness of its designers.[37]

The much maligned Challenger 1 proved to be an excellent combat vehicle in Operation Granby and it served throughout the 1990s. This example is the CO 1st RTR's mount photographed in 1996. (*Tim Neate*)

These Royal Dragoon Guards Challenger 1s are being prepared for a hard day's training in 1996 on the Salisbury Plain training area. (*Tim Neate*)

The Queen's Royal Hussars were created from the amalgamations that followed the end of the Cold War. The Queen's Royal Irish Hussars who had employed the Challenger 1 in the 1991 Gulf War and the Queen's Royal Hussars operated the Challenger 1 until the Challenger 2 was received in 1998. The regiment's Challenger 1s also served in Bosnia, although this example was photographed in the United Kingdom. The turret roof layout was very similar to that of the Chieftain that preceded it. (*Tim Neate*)

# Chapter Eight

# Challenger 2 Prototypes

Vickers Defence was obliged to meet measurable performance milestones throughout the early stage of the Challenger 2 programme. Reliability, battlefield survivability, and habitability were all established in gruelling trials. A total of 9 Challenger 2 prototypes were built as complete tanks fitted with turrets differing in some details. For example the original prototypes did not all include the L30 gunner's secondary telescopic sight (although it was included on all vehicles used in gunnery trials and retrofitted to the others).

Prototype V1 (06 SP 87) was employed as a general trials and demonstration vehicle including use in the environmental trials. This vehicle became the standard vehicle for *Project Copenhagen* – the modifications required for the Omani Challenger 2 order. It has since been used to represent the Challenger 2E on occasion. Prototypes V2 (06 SP 88), V3 (06 SP 89), and V4 (06 SP 90) were all used for Reliability Growth Trials, and later for most of the automotive mileage trials. Prototype V5 (06 SP 91) and Prototype V8 (06 SP 94) were employed for army user trials by the Armoured Demonstration and Trials Unit (ADTU Bovington), with V5 being mostly employed at RAC Gunnery School at Lulworth in Dorset. Prototypes V6 (06 SP 92) and V7 (06 SP 93) were both used in firing trials and weapon proving.

Prototype V9 (06 SP 95) was originally the Challenger 2 definition vehicle that was kept to the latest British build standard. It also served as a Vickers Defence sales and demonstration vehicle and it was marketed as the *Desert Defender* in Middle East in 1992. V9 was later used as the base vehicle for *Project Exmouth*, which resulted in its transformation into the Challenger 2E. This tank was later fitted with Production Turret No. 2 which received upgraded sights for the Hellenic Army trials (requiring an additional production turret to be ordered to fulfil the British Army production order). In addition two turrets were built exclusively for test purposes: Turret TA1 was employed for rig testing of the weapon systems and Turret TA2 underwent firing tests to prove its armour layout. Most of these vehicles have survived to the present day although some have become static exhibits or gate guardians.[38]

The Challenger 2 began production in 1993 at Newcastle, and the new tank was officially accepted by the army in May 1994. The Challenger 2 order was increased to 386 MBTs and another 9 Driver Training Tanks in July 1994 once the decision to sell the Challenger 1 fleet to Jordan was finalized. On 1 August 1994 the first production vehicle rolled off the assembly line, much to the satisfaction of the Vickers board of directors and of the Vickers and Ministry of Defence development teams. The Challenger 2 had been eight years in the making and was expected to enter service in 1995. By this circuitous route the British Army had its Chieftain replacement; the last Chieftain was retired by the 1st Royal Tank Regiment in late 1995.[39]

## Chapter Nine

# The Challenger 2 Described

The Challenger 2 rectified all of the reliability problems that had dogged the Challenger 1. The new David Brown TN54 automatic transmission gave it incredible agility for a 62-ton vehicle and its improved Hydrogas suspension allowed a much better cross country performance than given by its predecessor. The Challenger 1's CV12 was retained for the new tank. The full designation for the engine is the Perkins Engines Company CV12 TCA V-12 12-cylinder turbo charged (aspiration, air to air charge cooling) 26.1 litre diesel Model No.3 Mk.6A. It produces 1,200bhp at 2,300 revolutions per minute. The auxiliary generator was originally a Perkins P4108 but it was replaced after the first decade of service with the ESU Auxiliary Power Unit. The TN54 has 6 forward gears and 2 reverse gears, and can achieve neutral turns, a useful manoeuvre in confined areas. The gear box transmits power from the main engine through a splined coupling to a torque convertor, gear train and clutches to the output shafts. Gear selection via the driver's gear-range selector is connected electrically to the control system, enabling the driver to preselect six forward and two reverse gears to suit all driving conditions.

The Hydrogas system is easily maintained and repaired; a relatively small hydrogas unit is far easier to replace than the torsion bars used on other MBT suspension types. The Challenger 2 hull has six suspension units on each side, and each unit floats on a mixture of compressed oil and nitrogen gas. The units function in three phases. If the vehicle is static or is moving along a firm reasonably level road with minimal movement on the axle arm pressure on the hydraulic fluid will be low. If, however, the terrain causes large upward deflection of the wheel the piston moves, pressurizing the oil in the cylinder while a separator piston compresses the nitrogen. Once the upward force decreases the nitrogen gas expands and the reverse action takes place, a process that absorbs the shock of sudden or gradual wheel movement that is repeated over and over again. The system combines the smoothest of rides with a very stable firing platform and minimizes both wear and tear and crew fatigue. One strange aside is that when the hydrogas suspension system units become cold during prolonged periods of inactivity they tend to settle. For example on exercises in cold weather at daybreak the tank might be sitting at strange angles! Once running the units soon warm and the correct ride height is resumed.

Correct track adjustment reduces fuel consumption and prevents wear to the sprocket teeth. 'Track bashing' is a major task on a tracked vehicle of any size; this describes the maintenance of the tracks and correct track tension. It often requires the removal of track links which are designed for a specific mileage during which the track blocks and the connector pins are subjected to immense stresses. A well-used track pin resembles a cam shaft, worn by the constant movement, eventually causing stretching when measured collectively over the entire track and resulting in decreased tension. This problem of a slack track becoming easily thrown is overcome by the tension being adjusted by moving the idler wheel. On old fashioned idler-tensioned tracks with Horstmann (like the Chieftain) or Hydrogas suspensions (like the Challenger 1) the tank required constant tightening or slackening of the track.

On these older tanks types the track was adjusted by using a very large ratchet spanner to engage the track adjuster nut. On the Challenger 1 the driver would manoeuvre the tank to bring the slack track to the front of the vehicle, either by steering in neutral or by slowly driving forwards. With the spanner fitted to the track adjustment nut and a long extension bar attached for leverage, one crewman would tighten the track simply by raising and lowering the par, although as it tightened the efforts of two men standing on the extension bar were inevitably required (often jumping up and down which was dangerous in wet weather). For ease and safety the army introduced a hydraulic track adjuster system on the later Chieftain Bridge Layers; this was carried forward to the Challenger 2. Hydraulic track adjustment also allowed the track to be adjusted in a contaminated environment.

The hydraulic track adjustment system has two externally mounted double-action hydraulic rams linked to the front idlers by means of piston ram eyes. The rams are controlled by the driver from a control unit inside the driving compartment. Each unit has two levers marked with left, right, extend, neutral and retract. To use the system the auxiliary generator and main engine must be running and the slack track is brought to the top of the run by use of the steering levers. Once there the driver keeps the engine revolutions at around 1200rpm and applies the handbrake, after which he can select the side that needs adjusting and adjust the track to the required tension for each side. A manual pump is provided to enable the track to be adjusted if the power pump system fails. In the event of mechanical failure there is an emergency tensioning procedure provided by a special bolt fitted to the rear of the ram which winds it forward manually.

The Challenger 2 design employed a significant number of foreign components in its turret systems, a first for a British MBT design but very much in keeping with the Vickers design philosophy of putting reliability ahead of other considerations.[40] The original Mk.7 turret's fire control arrangement evolved considerably as new products became available for the production Challenger 2. The fire control computer was changed to the Canadian CDC type also used on the M1 Abrams; the commander's sight was the French SFIM panoramic type (SAGEM having absorbed SFIM) and the gunner's sight included Barr and Stroud and SFIM components but the best features of the Challenger 1 (like the TOGS thermal gunnery system) were also carried forward to the Challenger 2 in improved form. The Challenger 2's turret is operated by the commander, loader and gunner and is provided with an electric traverse system permitting a 360 degree traverse in 12 seconds. The turret houses the NBC filtration system, the main and secondary armaments, the fire control system and the tank's communication system. The level of electronic systems and mechanical reliability achieved with the prototypes was impressive and the crew feedback was enthusiastic at every level.[41]

The commander surveys the battlefield with the gyrostabilised, fully traversable SFIM VS 580-10 panoramic sight. The upper unit of the VS 580-10 houses the sight and electronics, and it is mounted in front of the commander's cupola. An Nd:YAG laser rangefinder is incorporated into an intermediate assembly underneath the cupola which joins the traversing sight head to the commander's viewing telescope inside the turret. The telescope includes the viewing system, hand controls, electronics and the sight stabilizer. The VS 580-10 traverses 360 degrees without any need for the commander to move his head, and the sight can elevate from +35 to −35 degrees depression to permit tracking of a moving target over the roughest terrain. The commander's sight can view at 3.2x magnification with a field of view of 16 degrees or at 10.5x magnification optics with a 5 degree field of view. The cupola incorporates 8 periscopic viewing devices to permit the commander to observe the tank's general surroundings or identify nearby targets.

The gunner is positioned immediately in front of the commander and in combat a good gunner is worth his weight in gold. The Gunner's Primary Sight (normally referred to as the GPS) consists of a sight body with a visual sighting channel and a sight head with a stabilized aiming mirror. The gunner aims the main and coaxial armament through a monocular eye piece. The GPS was developed by combining a SFIM-designed stabilized sight head with Barr and Stroud components to Vickers's requirements. The 4 Hz Nd:YAG laser rangefinder incorporated in the GPS has a range from 200 metres to 10 kilometres. The backup Nanoquest L30 direct sight telescope is mounted coaxially in the mantlet with an armoured cover. This simple sight allows the gunner a reversionary gunnery sight if the GPS is damaged in battle. Another viewing device is the Gunner's Unity Window which, using periscopic mirrors, allows the gunner a fixed view directly ahead of the tank (but for the use of mirrors, his position would offer nothing more than a good view of the rear face of the TOGS barbette!).

The Challenger 2 incorporates the Thermal Observation and Gunnery Sight II (known as the TOGS II from Pilkington Optronics). The unit is mounted on the upper face of the gun mantlet in an armoured barbette which houses the TICM 2 thermal imager (known as the TISH or Thermal Imager Sight Head), providing thermal vision in the line of sight of the main gun. The TISH is constantly cooled by a compressor and gas bottle. The TOGS system allows the thermal image of the target to be viewed in the sights of both commander and gunner; the aiming marks on the thermal image are overlaid and the TOGS image can function as the sight picture in the fire control system. The thermal image, which can be viewed at 4x magnification or at 11.5x magnification in the sights, is also shown on television style viewing monitors at the commander's and gunner's positions for general battlefield observation purposes. The Challenger 2's thermal viewing capabilities are excellent but only function in the line of sight of the main gun. The tank, unlike the Vickers Mk.7, currently lacks an independent thermal viewer visible in the commander's sight. This could be added when it becomes an operational requirement and is already featured in configurations proposed for the Challenger 2E.[42]

The production Challenger 2 took longer than expected to enter service. Three of the first six tanks from the production line failed quality tests in late 1994 proving that the standard of the production tanks did not meet that of the prototypes. As a result of this strict contract scrutiny by the Ministry of Defence the first production batch was rejected and had to be reworked by Vickers. This was the first time the MoD had held a manufacturer's feet to the fire. Remanufacture had a knock-on effect in combination with slow production and continuing evaluation and the in-service date for Challenger 2 slipped back. Reliability Growth

Test trials to ensure that all deficiencies were rectified in the first production batch were conducted in July 1996. The Ministry of Defence held Vickers very tightly to the terms of the contract and would not accept any tanks until all faults were rectified.[43]

Eventually the problems were resolved and in January 1998 the Royal Scots Dragoon Guards received the first of their tanks to become the first operational Challenger 2 regiment (with thirty-eight tanks) in June. Prior to the tanks being issued to the regiment, each passed through a vigorous commissioning process. All subsystems were tested in the hull and turret, and all weapons were live fire tested to identify any outstanding faults. For the first time a British armoured regiment received perfect equipment and any teething problems were minimal. The 2nd Royal Tank Regiment was next to be re-equipped in the second half of 1998. Within two years the Challenger 2 was the Royal Armoured Corps' standard MBT.[44]

# Chapter Ten

# Early Service

After the successful conclusion of Operation Granby, the Royal Armoured Corps passed through the hard amalgamations necessary to reduce the size of the army. The die was cast during the course of 1990 but the reductions took place in 1992-1993 by which time the Royal Armoured Corps had shrunk from nineteen to eleven regiments. Until the Challenger 2 became available in 1998 these regiments retained the Challenger 1. The next major government effort to reduce the forces came in 1998 with the Strategic Defence Review which coincided neatly with the Challenger 2's entry into service.[45]

The disappearance of old communist regimes created instability in the Balkans where civil war and genocide led to intervention by the United Nations (and subsequently by NATO). Once the Challenger 2 was operational it was a matter of time before the new British MBT deployed on operations. The Challenger 2's first such use was as part of the SFOR stabilisation force in Bosnia. By the time it was sent there the chances of the new tank engaging in combat were very remote. Ironically SFOR were keen to lower the profile of their patrols and a 62 ton MBT was probably not the best way of doing that. As the Bosnian situation slowly calmed in 2000 trouble flared up in Kosovo between Serbian forces and the Muslim population in yet another ethnic conflict. The Royal Scots Dragoon Guards (SCOTS DG) were deployed with NATO's KFOR to enforce peace.[46]

The SCOTS DG's organization for KFOR duties was not a typical conventional war deployment. Only B Squadron brought their Challenger 2s while the rest of the regiment was employed in a dismounted role – as was normal for peacekeeping duties. The small number of tanks symbolised KFOR's hopes of keeping a low profile while retaining the option to engage any hostile forces. B Squadron was used to patrol areas between the warring factions and to act as a highly visible deterrent. Some of the SCOTS DG Kosovo patrols travelled distances in the region of two hundred and fifty kilometres. The KFOR Challenger 2s carried the upgraded Chobham modular side armour and ROMOR reactive armour arrays on the hull front. Several vehicles were also fitted with the Pearson combat dozer blade (which proved useful for clearing road blocks). Rotations into the Kosovo peacekeeping force by the other Challenger 2 regiments followed as each worked up on the new tanks. In the entire deployment the Challengers saw no combat but demonstrated impressive reliability despite the weight of their increased armour.

As the Challenger 2 entered service in 1998 it was provisioned with a scale of spares that permitted complete vehicle availability. Eventually the optimum spares levels holdings were identified and were set out in the Challenger Innovative Spares Provision programme (CRISP). This programme was designed to keep costs to a minimum while maintaining an effective stock of spares to prevent shortages from impacting MBT availability as it had at the time of Operation Granby. In May 2000 a contract valued at approximately £120 million was awarded to Vickers Defence Systems (by then already bought out by Rolls Royce PLC) and the Ministry of Defence's newly formed Defence Logistics Organisation (DLO). The contract was issued for the shared base inspection and repair programme for the Challenger 2 and its associated Driver Training Tank. In January 2001 Vickers Defence Systems partnered with the Ministry of Defence's ABRO maintenance agency to provide integrated maintenance and repair service to the Challenger 2 fleet under contract to the Defence Logistics Organisation. It was expected that around 5 per cent of the total Challenger 2 fleet would be undergoing a complete 5 month strip to components level overhaul at any one time in ABRO facilities. Repairs and preventive maintenance could be carried out by Vickers and ABRO technical personnel as particular circumstances required.[47]

A KFOR Challenger 2. The Royal Scots Dragoon Guards (SCOTS DG) were the first Challenger 2 equipped regiment and were also the first to deploy the new tank on operations. Note how this vehicle is equipped with ROMOR explosive reactive armour on the lower front plate and Chobham arrays on each side. (*Published by the Ministry of Defence © Crown Copyright 2009. Reproduced under Open Government Licence*)

The Challenger 2 was used for patrolling and to enforce NATO's presence in Kosovo. (*Published by the Ministry of Defence © Crown Copyright 2009. Reproduced under Open Government Licence*)

Second generation Chobham armour (known as Dorchester), an improved 120mm rifled gun and a fine hunter-killer fire control system were the new features introduced with the Challenger 2. (*Peter Brown*)

As delivered, the Challenger 2 weighed over 62 tons. Production took place between 1994 and 2002 at Vickers. (*Peter Brown*)

This shows the armoured door that protects the gunner's primary sight in the open position, and the location of the TOGS system's thermal imaging head (TISH) which is mounted centrally on the mantlet above the L30A1 gun mounting. (*MP Robinson Collection*)

This 2000 photo shows an early Challenger 2's L8 smoke grenade dischargers; note how the L30A1's thermal sleeve is fitted and the first style loader's L37A2 GPMG mounting. (*MP Robinson Collection*)

This Challenger 2 from Badger Squadron, 2nd Royal Tank Regiment photographed in 2005 shows the fire simulation equipment fitted and the commander's panoramic sight traversed. (*Tim Neate*)

The driver's position is located centrally in the front of the hull. The single periscope is located directly behind the lift and swing type hatch. (*Pierre Delattre*)

The new double pin track type and Teutonic looking skirting plates were a big departure from the Challenger 1's look, though the earlier tracks and skirting plates were featured on the Omani Army's Challenger 2s. This early Challenger 2 has the original loader's GPMG mounting type that was quickly replaced in service. (*Peter Brown*)

The Challenger 2's hull resembled that of its predecessor externally but was completely different internally. The Challenger 2's hydrogas suspension and powertrain were improved versions of the types employed in the Challenger 1. *(Peter Brown)*

A rear view of a 1st RTR Challenger 2 photographed in 1999 during Exercise *Brave Guardian*. *(Tim Neate)*

The Challenger 2 can lay a smoke screen by injecting diesel fuel into its exhaust, an old Soviet trick dating back to the 1940s! He we can see the concept aptly demonstrated by *Ace* of the 1st RTR. (*Published by the Ministry of Defence © Crown Copyright 2012. Reproduced under Open Government License*)

A 1st RTR Challenger 2 and Fuchs NBC vehicle during a break in activities. Cross country operations, especially in damp weather, could lead to a quick accumulation of hundreds of kilograms of dirt. (*Tim Neate*)

A Challenger 2 of the King's Royal Hussars in 2001 marked with OPFOR crosses and carrying fire simulator laser projector and sensors. Simulators are an extremely important part of crew training and their use extends from the classroom to the manoeuvre area. (Tim Neate)

A view of the opposite side of the turret showing the receptor sensor and the wires strung beneath the gun barrel. This KRH Challenger 2 was taking part in Exercise *Wessex Warrior* in 2001. (Tim Neate)

A Challenger 2 of Egypt squadron, 2nd Royal Tank Regiment on manoeuvres on the Salisbury Plain Training Area in 2002. A year later the regiment was one of those that gave the Challenger 2 its baptism of fire during the invasion of Iraq. (*Tim Neate*)

While the temperate paint scheme worn by the example seen here was not that worn in Operation Telic, the ROMOR and Dorchester armour arrangement here is representative. This vehicle was photographed in 2003 at Bovington. (*Tim Neate*)

A close-up of the ROMOR explosive reactive armour that protected the lower front plate. The glacis plate itself was fitted with Chobham armour, but like the Challenger 1, the Challenger 2 was designed to fight from hull down positions and the lower front hull plate was made of conventional armour plate. (*Tim Neate*)

We can see here how well the ROMOR array protected the lower front plate. The ROMOR armour kits provided for the Challenger 2 in 2003 were from the same stock issued in 1991 for Operation Granby. (*Tim Neate*)

The major difference seen in 2003 was the provision of dust skirts, a lesson learnt during *Saif Sareea II*. (*Tim Neate*)

A KRH Challenger in 2003 in the course of preparation for a tow by the squadron CRARRV recovery vehicle. The large stowage bins that are fitted to the rear of the turret can be swung away from the turret walls to permit access to the engine deck with the turret traversed. (*Tim Neate*)

The Challenger 2 Driver Training Tank (DTT) is used to train tank drivers. The basic idea was tried and tested with the Challenger 1 DTT. This example was photographed at Bovington in 2015. (*Dick Taylor*)

Tank crews of the RTR bombing up, the corporal in the front of the picture is holding a HESH/PRAC practice round, which is a training round that simulates HESH in weight and flight, but contains no explosive. The crew on the tank are also loading Practice APFSDS rounds (known as 'Fin' rounds). Both HESH/PRAC and practice 'Fin' can be identified as practice ammunition by the NATO designated colour of Saxe-Cobourg blue. Sharp eyes will pick out the Australian uniform of an RAAC exchange officer. (*Crown Copyright MoD*)

A Queen's Royal Lancers Challenger 2 being prepared for a pack lift in 2004. (*Tim Neate*)

A Foden heavy wrecker and the CRARRV are in attendance and can change the engine in under an hour. (*Tim Neate*)

The Challenger 2's power pack can be removed and changed as a complete assembly as seen here. The engine and transmission assemblies can also be removed separately if required. (*Tim Neate*)

The CRARRV was based on the Challenger 1 and first saw action in Operation Granby in 1991. This vehicle was photographed in 1999. The CRARRV has since been modernised with new tracks to serve alongside the Challenger 2 and can be fitted with appliqué armour as required. (*Tim Neate*)

# Chapter Eleven

# *Saif Sareea II*

The 1998 Strategic Defence Review cut the RAC again, and the number of MBT-equipped armoured regiments was reduced to six, but the size of the Challenger 2 regiment grew from thirty-eight vehicles in three squadrons of twelve MBTs (settled on by 'Options for Change' in 1990) up to fifty eight tanks per regiment with four sabre squadrons. The six regiments were The Queen's Royal Hussars, The King's Royal Hussars, The Queen's Royal Lancers, The Royal Dragoon Guards, the 2nd Royal Tank Regiment and a single squadron of the 1st Royal Tank Regiment. These were known as Type 58 regiments, although some operated with only three MBT squadrons, with forty-three tanks. The British armoured force was to be deployed in three armoured brigades (the 4th, 7th and 20th) in the 1st (UK) Armoured Division and the 3rd (UK) Mechanised Division. Each armoured brigade was made up of two Type-58 regiments and a mechanised infantry regiment with Warrior mechanised infantry combat vehicles. This establishment was leaner than the Cold War era armoured brigades and had been fully implemented at the end of the century. It was envisioned to be the most flexible and potent armoured formation ever deployed by the British Army. Great emphasis was placed on battlegroup operations and on the principle of all-arms cooperation. It was with this organisation that the Challenger 2 saw battle.[48]

In Iraq Saddam Hussein's regime still refused to comply with United Nations resolutions in 2001 and the remainder of the Middle East was far from peaceful. The British government decided to run a 22,000 man, two month exercise in the area called Exercise *Saif Sareea II* in the second half of the year. Given the close military alliance between the United Kingdom and United States, *Saif Sareea II* was no doubt intended to remind potential opponents that the British could deploy conventional forces to the area if required, and deployment strategies and desert tactics were extensively studied. *Saif Sareea II* was conducted in Oman and the troops worked closely with the Sultan of Oman's armed forces (themselves users of the Challenger 2). Ignoring local advice, the British decided that they would operate in the most inhospitable area of northern Oman, where even the Omani Army did not bother to train. The 4th Armoured Brigade sent some sixty-six Challenger 2s (with four squadrons from the Royal Dragoon Guards and a fifth from the Queen's Royal Lancers) to Oman, and Omani forces also took part. No extra money was allocated for desert modifications for the exercise because the British government did not see the need, so the Challenger 2 configured to fight in Europe participated in the exercise with predictable results. Oman is one of the dustiest and dry areas on Earth, and with no special sand filters or other theatre-specific equipment the Challenger 2's vulnerability to conditions in the desert was evident. Given the experience of operations in Kuwait and Iraq in 1991 the lack of mechanical precautions organised at the beginning of *Saif Sareea II* is difficult to understand.[49]

After the troops had acclimatised to Oman (with temperatures reaching 54 Centigrade) the exercise started and almost immediately problems arose with air filters that were supposed to last at least twelve months in the European theatre. The fine sand in northern Oman necessitated daily, and often hourly, filter changes. A sharp rise in the operation's costs quickly occurred as each filter cost £1,000 and sixty-six tanks were using them daily. On the final day only two RDG squadrons were fit for duty. As a stop gap each squadron stood a troop down and QRL were sent home to help reduce costs. With all attention being focused on keeping the Challenger 2s equipped with airlifted spares the soft skinned support vehicles were soon being neglected. Many of these broke down and it became painfully apparent why the Royal Omani Army did not train in that part of the country.

As so often stories soon appeared in the British media portraying *Saif Sareea II* as a major condemnation of the effectiveness of the Challenger 2 with suggestions that the MoD had got it wrong again and had bought a white elephant. The extent of the press criticism of the army and its weapons resulted in a parliamentary enquiry to investigate the issues.[50]

The enquiry found that the tank itself was not at fault, but rather that the desk warriors were again to blame for not scaling enough spare parts (and for picking a uniquely unsuitable exercise area).

'The Department's planning of the Exercise was deficient. In particular, because of poor information and inadequate analysis, it did not modify the Challenger 2 Main Battle Tank to operate in desert conditions.'[51]

Improvements in the scale of parts and in the quality of specialised components like tropical air filters were immediately put in hand to allow Challenger to fight in desert conditions. If the experience was slightly embarrassing for the army, the exercise had shown what needed to be done to make the Challenger 2 an all-round performer – and to that end *Saif Sareea II* was at least worthwhile. After the chastening lessons of the desert the army was very soon preparing again for the Challenger 2's next deployment.[52]

Exercise *Saif Sareea II*, Camp South, Oman, in September 2001. Challenger 2s of the 3rd Troop, D Squadron, The Royal Dragoon Guards seen advancing through the Thumbrait training area in Oman. (*Published by the Ministry of Defence © Crown Copyright 2015. Reproduced under Open Government Licence*)

*Saif Sareea II* contributed to the Challenger 2's excellent performance in 2003. (*Published by the Ministry of Defence © Crown Copyright 2001. Reproduced under Open Government Licence*)

D Squadron, Royal Dragoon Guards formed the core element of the 4th Armoured Brigade during *Saif Sareea II*. (*Published by the Ministry of Defence © Crown Copyright 2001. Reproduced under Open Government Licence*)

The RDG and the Queen's Royal Lancers participated amongst 20,000 British troops deployed from the UK and Germany to train alongside Omani forces in *Saif Sareea II*. (*Published by the Ministry of Defence © Crown Copyright 2001. Reproduced under Open Government Licence*)

The choice of the Hughes Chain gun as the coaxial weapon for the Challenger 2 was made on the premise of higher reliability due to the fact that this weapon was electrically operated (rather than gas operated like the GPMG coaxial weapon on the Chieftain and Challenger). This weapon was also used on the Vickers Mk.7 MBT. (*Lawrence Skuse*)

The glacis and turret front are the most heavily armoured parts of the Challenger 2. Its Chobham armour and the tank's stock armour layout reflects the fact it was originally designed to fight from hull down positions against Soviet MBTs like the T-80 and upgraded T-72s. The Challenger 2's armour protection has since been revised with modular armour to meet the threats posed by Improvised Explosive Devices, advanced Rocket Propelled Grenade (RPG) types and new Antitank Guided Weapons. (*Lawrence Skuse*)

The growth of NATO into Eastern Europe was a new phenomenon in the late 1990s. These British officers show visitors from the Czech army a Challenger 2 named *Carefree* at Sennelager in 2002. Vickers then was still hoping to sell the Challenger 2 to foreign customers. The few third generation MBTs sold in Eastern Europe were inevitably used Leopard 2A4s, partly because Poland and the Czech Republic had their own tank manufacturing capabilities. (*Lawrence Skuse*)

# Chapter Twelve

# Operation Telic

The Challenger 2 made its well-publicised combat debut when the British government supported the controversial American decision to invade Iraq in 2003. In terms of boots on the ground, the army committed an armoured division, albeit on a smaller scale than in 1991. The 1st (UK) Armoured Division included the 7th Armoured Brigade, the 3rd Commando Brigade and the 16th Airlanding Brigade in its order of battle during Operation Telic. The 7th Armoured Brigade was commanded by Brigadier Graham Binns throughout the operation, but the tone of the operation was very different from Operation Granby's blitzkrieg style of war. During Operation Telic the 7th Armoured Brigade was formed into four armoured battlegroups composed of mixed armoured squadrons and mechanized infantry companies. The brigade operated in two armour-heavy and two infantry-heavy groups; this order of battle was conceived to fight a mobile battle deep in the enemy's territory with artillery support from a regiment of thirty two AS-90 self-propelled guns and MLRS batteries.[53]

The SCOTS DG Battlegroup was formed from two squadrons of the Royal Scots Dragoon Guards (SCOTS DG) and the Irish Guards with forty-two Challenger 2s and twenty-eight Warriors (later a squadron of Challengers was detached to support the 3rd Commando Brigade). The 2 RTR Battlegroup was formed from two reinforced 2nd Royal Tank Regiment squadrons with forty-two Challenger 2s and twenty-eight warriors from The Light Infantry. The Black Watch Battlegroup was formed from a detached squadron each of the 2nd RTR and SCOTS DG with twenty-eight Challenger 2s and forty-two Warriors from the Black Watch. Finally the Royal Regiment of Fusiliers Battlegroup was formed from eighteen Challenger 2s of the Queen's Royal Lancers and forty-two Warriors of the Royal Regiment of Fusiliers. The armoured regiments employed experienced soldiers wherever they could and tank troops were largely commanded by full lieutenants and captains.[54]

The 1st (UK) Armoured Division deployed 116 Challenger 2 gun tanks in all. The armoured regiments' pre-deployment training, like that of the associated infantry, had reflected the expectation of fighting another Operation Granby. The reality of Operation Telic was different and unexpected. Events before the war were familiar and some men recalled those of a dozen years earlier (the Scots DG actually started Telic from the very locale where they had ended Granby!). After moving to Kuwait in-theatre training was completed before the tanks were moved for upgrading to an area codenamed Camp Coyote. Up-armouring kits were fitted and special desert modifications were completed. Finally the tanks were bombed up with full ammunition loads and the men awaited the order to advance.[55]

The ground war started on 21 March 2003 and the British advanced towards the city of Basra a day after the air campaign began. While the force was generally well prepared, some supply items were still of concern. The British division advanced about a hundred kilometres on the first day and met little opposition. It was quickly noticed that they had to advance through abandoned positions and move deep into Iraq to find the enemy. During the advance the T55s and T59s concealed in fortified positions were all abandoned as the national army disintegrated; this changed on ground that the Fedayeen political militia decided to hold. Once found, they were willing to fight and their training in camouflage and concealment proved to be very good. Enemy tanks consisting mostly of T55s were deployed in tank scrapes.[56]

The role of the Challenger 2 switched almost from the second day of the advance from that of the armoured spearhead at the tip of each battlegroup into one of infantry support or of leading armoured raiding columns. This was the start of a hard stage of the war for the soldiers, playing a waiting game against irregulars and political troops, had to be mindful of causing civilian casualties. One of the 2nd RTR's first heavy contacts with enemy forces in this role took place at Az Zubyar. The original plan called for teams of snipers to take up positions overlooking a Fedayeen headquarters ready to eliminate anyone still there. Unfortunately the snipers were spotted and their intended targets slipped away. The Fedayeen later engaged the tanks and infantry with rocket propelled grenades (RPG) and heavy machine-gun fire. The 2nd RTR tanks and the Warriors of the accompanying infantry responded with 7.62mm chain gun fire and 120mm HESH fire before pulling back and taking the sniper team with them. The road to Basra lay open but the decision to surround the city was taken and the tempo slowed.[57]

The last stages of the advance on Basra were slow, frustrating and consequently cautious; during this period tragedy struck. Even with thermal imaging equipment nocturnal operations are confusing and the risk from friendly fire is much higher than in daylight. In a night engagement against Iraqi armoured forces on 25 March a Challenger 2 belonging to the Queens Royal Lancers was mistaken for an enemy vehicle. It was fired upon and destroyed by a HESH round fired by a 2nd RTR Challenger 2. Details of the inquiry that followed showed that, despite provision of IFF (Identification Friend or Foe) panels developed from experience in Desert Storm and the use of TOGS, the QRL Challenger 2 had been mistaken at long distance for an Iraqi MTLB operating in the same area. The HESH strike had freakishly passed through the open commander's cupola hatch and the tank exploded as it detonated inside the turret. The ammunition inside the tank exploded through sympathetic detonation even though the bag charges were inside armoured bins. The explosion lifted the turret into the air and dropped it onto the engine deck; the vehicle commander, Corporal Steven Allbutt, and the driver, Trooper David Clarke, were killed. The gunner and loader were badly injured by the explosion although they were standing outside the tank. This was a tragic incident of friendly fire and, although blame was attached at the subsequent inquiry, it was an accident of war.[58]

This incident immediately resulted in inevitable criticism from those who argued that if Challenger 2 had been fitted with ammunition blow off panels like the Abrams it would not have exploded. This criticism is baseless because it ignores how ammunition is stored in the Challenger 2's interior. Due to the very freak nature of the hit (which probably could not be replicated) the verdict is that, while closer adhesion to standard operating procedures might have avoided the misidentification, blow off panels in the turret roof would have made no difference.[59]

The battles around Basra raged under direct media coverage; one of the most publicised engagements was carried out under command of Major Tim Brown of A Squadron, SCOTS DG on 29 March 29. The target of one of the armoured raids was a Fedayeen base equipped with a communications tower used for propaganda broadcasts by the Ba'ath party. The tower was knocked out with an APFSDS round fired by Brown's gunner Corporal Vince McLeod. During the shoot the gunner actually used the muzzle reference system to check calibration. This feat of gunnery equalled the Challenger 1's long range kill a few years before and deprived the Ba'ath party of its propaganda weapon. The entire engagement was filmed by an embedded BBC team and appeared on live television in Britain. The operation proved to be something of a media hit.[60]

More small pockets of resistance were met but Basra was completely surrounded by coalition forces. There was no wish to make a Stalingrad out of Basra and every attempt was made to use the available military intelligence in the planning of local operations instead of simply smashing forward into the city. The tactics of Saddam's loyalists were completely different from those of the Iraqi army in 1991, possibly because they were defending their own country and partly out of the sort of desperation that one sees in men faced with the loss of power and status. Coalition forces had to be extremely careful about casualties and political support at home for Operation Telic was tenuous.

The main threat to armour was the rocket propelled grenade, and every armoured squadron sustained multiple RPG attacks during the battles around Basra. The Challenger 2s of the 2nd RTR were repeatedly hit by RPG projectiles but without serious damage (most commonly their optics were impaired). Fedayeen antitank teams were shot down with co-axial machine guns or blown to pieces with HESH. The Challenger's Chobham armour proved invulnerable to RPG fire; one tank was recorded to have been hit 70 times without serious damage. As an antipersonnel weapon 120mm HESH proved lethal at ranges as short as 25 metres. The new CHARM depleted uranium APFSDS-T ammunition was never used against state of the art enemy main battle tanks but many rounds were fired at both enemy strongpoints and pinpoint targets.[61]

Many battles were fought before the final cease-fire on 8 April but one particular incident demonstrates the protection provided by the Challenger 2. This concerned a tank from C Squadron SCOTS DG which, due to errors from the commander and sheer bad luck, was destined to go into regimental lore. C Squadron was assigned reserve duty on 23 March and camped on Shaibah airfield for several days while their comrades in arms were heavily committed. The 3rd Commando Brigade requested their assistance in clearing the Al-Faw peninsula. Advancing to contact with the 3rd Commando Brigade meant crossing enemy held territory and a small enemy armoured force was eliminated *en route*. On 29 March the squadron deployed in support of the Royal Marines to capture the Abu-al-Khasib area and outlying settlements. Callsign 21 was manoeuvring along a causeway in the marshlands of Abu-al Khasib, south of Basra, when it came under sustained fire including RPG fire aimed at the vehicle optics. One RPG round destroyed the driver's periscope leaving him blind. The Fedayeen employed human shields, including children, to prevent the British from returning fire.[62]

Callsign 21's crew could not return fire for fear of killing civilians, but once the children were freed the tank opened up with HESH and the coaxial chain gun, killing several of the enemy whose fire became more intense. The crew tried to reverse around a parked British truck, but the unfortunate driver (who was reversing under the commander's instructions) pulled the wrong steering lever resulting in the tank sliding off the causeway and shedding both tracks. The Iraqis saw the tank's predicament and intensified their fire skilfully attacking the tank's vision devices. The gunner's primary sight and two of the commander's cupola periscopes were destroyed. The turret was next hit by a Milan wire-guided anti-tank guided missile, but again the Chobham armour held out (although the concussion caused the loader a broken wrist).

Inside the stranded tank its commander knew that he and his men could not fight back. Recovery was arranged by radio, and the first CRARRV recovery vehicle turned up to try and winch CS21 out. The combination of the angle of approach and the broken track jammed around the sprocket made recovery an impossible task, and in the end the winch failed. A second CRARRV managed to pull the stricken tank onto the causeway but the angle was too steep, and CS21 slid back into an even worse position (and the winch rope on the second CRARRV failed). It seemed that the tank would have to be abandoned and destroyed by an air strike. Luckily the REME detachment had not given up and, with the squadron commander laying down suppressive fire, the two CRARRVs were hooked together with recovery bars. Finally they hooked up the tank and managed to recover it – a feat that had taken fourteen hours. While the damage to CS21 sounded severe, it is recorded that within four days the damaged optics were replaced with parts sent from Britain and the tank was back in action. If nothing else this incident showed that the Challenger 2 was living up to its reputation for protection.[63]

The battle groups of the 1st Armoured Division carried out constant patrols along the highway leading into Basra but the Fedayeen fought some very skilful actions, and did not seem to wilt in face of the armoured assault. During one such patrol conducted on 6 April by Major Brannigan's B Squadron, SCOTS DG up the Basra highway sector codenamed Red Six, a seemingly abandoned T55 suddenly let loose a 100mm round at point blank range at the lead Challenger 2 (commanded by the troop leader Captain De Silva). The round hit directly in the middle of the ROMOR explosive reactive armour pack on the glacis, detonating two reactive armour bricks. The ensuing massive explosion caused the rest of the troop to fear the worst but it was just the ROMOR doing its job. The offending T55 was engaged by the whole troop and was shot to pieces.[64]

The 7th Armoured Brigade (which included the 2nd Royal Tank Regiment and the SCOTS DG) prepares to deploy to the Gulf in early 2003. A quick range period firing live ammunition was conducted on the Hohne Ranges in January 2003 prior to deploying on Operation Telic. (*Published by the Ministry of Defence © Crown Copyright 2003. Reproduced under Open Government Licence*)

A Challenger 2 Main Battle Tank from the Royal Scots Dragoon Guards travels along the road to the firing range at Bergen Hohne Training Ground in Germany. The 7th Armoured Brigade (The Desert Rats) was based in Germany and was ordered to move to Kuwait between the last week of February and the first week of March 2003. (*Published by the Ministry of Defence © Crown Copyright 2003. Reproduced under Open Government Licence*)

A Queens Royal Lancers Challenger 2 tank, attached to the 1st Royal Regiment of Fusiliers (1 RRF) Battlegroup, demonstrates its agility to the world's media after up-armouring at Camp Coyote, Kuwait, in early March 2003. The armour upgrades and desert modifications were made at Camp Coyote and caused some disruption to planned last minute infantry exercises by the 3rd Commando Brigade. (*Published by the Ministry of Defence © Crown Copyright 2003. Reproduced under Open Government Licence*)

A Challenger 2 main battle tank of the Queens Royal Lancers crosses an Iraqi defensive ditch by means of a General Support Bridge prepared by 39 Squadron, 32 Regiment, Royal Engineers. The QRL sent two squadrons to fight alongside the 2nd RTR and SCOTS DG in Operation Telic, losing one Challenger 2 in a friendly fire incident. The QRL took on the formation reconnaissance role in 2005, relinquishing the Challenger 2 for lighter AFVs. (*Published by the Ministry of Defence © Crown Copyright 2003. Reproduced under Open Government Licence*)

A Challenger 2 of the Queens Royal Lancers on Operation Telic. This tank's ROMOR reactive armour bricks were fitted still in their original factory NATO green. Note that the identification panels usually fitted to the turret are not yet fitted to this tank, although the mounting points are visible. (*Published by the Ministry of Defence © Crown Copyright 2003. Reproduced under Open Government Licence*)

The same Challenger 2 of the QRL in a close up, showing the ROMOR reactive armour array carried in a stoutly built frame on the glacis plate. On the side of the TOGS 2 system's TISH housing we can see the modern incarnation of the famous Death or Glory emblem of the 17th/21st Lancers superimposed on the crossed lances of the 16th/5th Queen's Royal Lancers, the two cavalry regiments from which the regiment was formed. (*Published by the Ministry of Defence © Crown Copyright 2003. Reproduced under Open Government Licence*)

Seen after Basra fell, a Challenger 2 of B Squadron, the Queen's Royal Lancers stands guard as Iraqi civilians try to get on with day to day life. We can see the dust skirts fitted at Camp Coyote have disappeared and that the tank carries several QRL decals on the turret sides and on the TISH barbette door. (*Crown Copyright, by kind permission of The Queen's Royal Lancers Museum*)

A somewhat battered QRL Challenger 2 in Basra. (*Crown Copyright, by kind permission of The Queen's Royal Lancers Museum*)

The Operation Telic operational medal, awarded to all British soldiers, airmen and sailors who served on any of the Operation Telic operations. (*Published by the Ministry of Defence © Crown Copyright 2015. Reproduced under Open Government Licence*)

A Challenger from the Royal Scots Dragoon Guards stands guard on the one of the main roads in Basra. During the invasion the SCOTS DG had squadrons detached to the 3rd Commando Brigade and the Black Watch Battlegroup, as well as forming the hard core of the SCOTS DG Battlegroup with the Irish Guards. (*Published by the Ministry of Defence © Crown Copyright 2003. Reproduced under Open Government Licence*)

The 2nd Royal Tank Regiment first received the Challenger 2 in 1998 and played a key role in the 7th Armoured Brigade's battle for Basra. This tank was possibly photographed at Camp Coyote in Kuwait before the start of combat operation in March 2003. Of note we can see the unusual eye design painted on the TOGS barbette beneath the 7th Armoured Brigade red jerboa. Note how the smoke dischargers have been painted sand and how the paint on the mantlet beneath the chain gun has worn off. Of the three regiments that took part in the invasion of Iraq, the 2nd RTR was the last to serve on MBTs; it exists today as The Royal Tank Regiment. (*Crown Copyright MoD, Courtesy Royal Tank Regiment*)

A Challenger 2 of the 1st Royal Tank Regiment on the training ground near Warminster in 2005. The 1st and 4th Royal Tank Regiments amalgamated in 1993, and the Scottish identity of the 4th RTR was preserved in the new regiment, along with the more official and very famous Chinese Eye markings. These 1st Royal Tank Regiment OPFOR Challenger 2s bore the Scottish Saltire unofficially on the TISH housing above the main armament. The Challenger 2 turret design featured a wide mantlet inherited from the Vickers Mk.7, which offered a more stable gun mounting than the narrow trunnions of the Chieftain and Challenger 1 allowed. This configuration allowed the Thales (formerly Pilkington) Optronics TOGS 2 system's Thermal Imaging Sensory Head (TISH) to be mounted on top of the mantlet, which addressed calibration issues inherent in the offset TOGS mounting on the Challenger 1. *(Lawrence Skuse)*

This Challenger 2 crew manoeuvres on the muddy training ground scanning the available targets. We can see an old Chieftain hard target in the background. (*Lawrence Skuse*)

The Challenger 2 received new lighter pattern road wheels after the invasion of Iraq. This 2nd RTR Challenger 2 was photographed on Salisbury Plain in 2005. (*Tim Neate*)

The use of identification panels in Operation Telic in 2003 (by US and British forces) extended rapidly to their use in training in the United Kingdom. They were carried on each side of the turret's front, and the rear. This Challenger 2 named *Abdullah* was photographed during *Exercise Urgent Quest*, with A Sqn, 1st RTR in 2005. (*Tim Neate*)

This Challenger 2 of the Queen's Royal Hussars, seen at Larkhill in 2004, still carries the appliqué armour around the driver's position. This was part of the original appliqué armour array specified to protect the Challenger 2's hull (composed of passive Chobham armour and reactive ROMOR ERA). This armour upgrade was inherited from the Challenger 1's *Operation Granby* configuration. It was re-issued and reused for the intervention in the Balkans, and for the 2003 invasion of Iraq. (*Tim Neate*)

# Chapter Thirteen

# Occupation and Urban Combat in Iraq

Telic was not a war like Operation Granby, and the Challenger 2 was in for a hard slog on occupation duties. The Challenger 2 was designed to take the punishment expected in fighting Soviet MBTs in the European theatre and to be upgraded with modular armour packs to meet increased or theatre specific threats. Modern MBTs carry most of their protection on the front face of the turret and on the glacis plate. Areas like the sides and the lower hull plates are not armoured to the same degree and rely on mission-specific appliqué armour. To provide maximum protection to the crew on operations in Iraq in 2003, the same modular armour packages employed in 1991 were carried again, though in modernised form (and with dust suppression skirts).[65]

When the ground war ended the Iraqi army was formally disbanded, and the 1st (UK) Armoured Division handed the Basra area over to the 3rd (UK) Mechanised Division. The continuing presence of foreign troops became a source of discontent to the Iraqis many of whom joined anti-government insurgent groups. These guerrillas soon created major problems and, as they obtained increasingly better weapons, the role of the occupation force in Basra became progressively more difficult.[66]

Occupation duties in Iraq presented urban warfare conditions for which the Challenger 2 had never been designed. All-around protection had to be provided against RPG attack with appliqué armour blocks added to the turret sides and with bar armour on the rear of the tank. The design was nothing more than sets of steel slats welded together to form a protective cage around the vehicle. The intention was to trap the warhead of an RPG round and possibly even detonate it before it could hit the armour of the hull or turret. This clearly succeeded as several vehicles were recorded as returning from patrols with RPG projectiles stuck between the slats. The Challenger 2's protection was constantly upgraded over the following years to meet the changing threat.[67]

The ROMOR ERA package was the first of the Challenger 2's armour components to be singled out for replacement. During the continuing Operation Telic patrols during August 2006 the driver of a Queen's Royal Hussars Challenger 2, Trooper Sean Chance, lost three toes when an RPG-29 tandem warhead round penetrated the explosive reactive armour protecting the lower bow plate of the hull during an engagement in Al-Amara. The tank had already been hit by ten to fifteen RPGs and was under sniper fire. Despite his injuries, Chance was able to reverse the vehicle to the regimental aid post. This incident caused great concern because the ensuing investigation concluded that the ROMOR ERA panels struck by the RPG had failed to explode. Less than a year later in April 2007 a large improvised explosive device (IED) shaped charge detonated underneath a Challenger 2 in Basra and penetrated the belly armour; the driver lost a leg and another soldier received minor injuries. The belly plate was a known weak spot on all of western MBTs against large IEDs; armour can be added to meet a threat to a particular area but the eventual result of too much would be an immobile hundred ton behemoth. Naturally this point was lost on the British media, who launched into another ill-informed sensationalist crusade blaming the MoD for knowing about this weakness and doing nothing. A comprehensive upgrade of the Challenger's frontal armour resulted, and the Theatre Entry Standard (TES) configuration was adopted (which gives maximum crew protection to vehicles deployed on operations and which introduced the new Dorchester block on the lower front plate).[68]

When the Royal Dragoon Guards returned to Basra for Operation Telic 11 they found that the Challenger 2s looked very different from their previous experience in theatre during Telic 5. The many precautions required against urban IED and RPG attacks led to the adoption of new side armour arrays for the hull; the turret sides were fitted with blocks of appliqué armour in addition to the bar armour described. The protection extended to electronic countermeasures equipment (to combat remote detonation devices) which was most visible on the rear of the Challenger's turret roof. The long flat antenna behind the crew's hatches was nicknamed the 'bird table'. As part of countermeasures to the IED threat, the 'bird table' and its associated antenna jammed electronic signals and equipment located on the front wings of the tank was designed to counter IED's.[69]

In the last years of the Basra deployments some of the tank squadron's Challengers had the loaders periscope from the turret roof replaced by a fully powered Telex remote weapon station (RWS). The RWS could be controlled and fired from under armour by the loader. The weapon station was fitted with thermal sights for full night fighting capability and it could mount either the M2 Browning HMG, a 40mm GMG, or the 7.62mm GPMG (although normally the latter was fitted). These heavily modified Challengers were nicknamed 'Street Fighters'; the same configuration can be fitted to any Challenger as required in the form of kits although mounting points, power lines and brackets have to be installed for the various components parts. Thus equipped, The Royal Dragoon Guards were heavily involved in the advance to re-impose British occupation on Basra in March 2008 at the head of the Scots Guards Battle Group supporting the Iraqi army's 'Operation Charge of the Knights' against the Mahdi's forces during Operation Telic 11. Similar measures to adapt the M1A2, Leclerc and Leopard 2 to urban warfare conditions have been taken to keep abreast of the threats associated with fighting irregulars in built up areas.[70]

The changing face of the Challenger 2's role in Iraq is also shown by the scale of deployments between the original regiments that spearheaded the invasion of Iraq in 2003 and the single Challenger-equipped squadron serving in 2006 on garrison duties at Shaibah airfield in the Basra area. After Operation Telic 1 an armoured regiment was always part of the Basra area garrison, but only one squadron was equipped with Challenger 2s while the remainder of the armoured regiment deployed in lighter vehicles.[71]

The Abrams performed impressively in Operation Iraqi Freedom (which was known as Operation Telic to the British) just as it had in 1991. Like the Challenger 2, the Abrams received numerous modifications in Iraq to improve its protection in an urban combat environment. (*U.S. Department of Defense*)

A Challenger 2 with bar armour and with reinforced turret sides being tested on the Bovington training area. While the basic Challenger 2 could carry up to 15 tons in additional armour kits, wear on road wheels increased and cross country mobility was somewhat reduced. These were necessary measures to maximise protection against RPG type weapons on operations in Iraq. (*Tim Neate*)

Late 2004 saw the first change in the Challenger 2's armour protection in Iraq. The RDG Challenger seen here named *Cruel Dragoon* carries bar armour on the hull rear and Chobham armour appliqué blocks on the turret sides. These first improvements were designed to increase the Challenger's protection from hollow charge warheads on areas not fitted with Chobham armour. It was the first stage in a deadly race between insurgent warheads and the Challenger 2's armour protection that was pursued for the rest of Operation Telic. The stop sign on the turret side is secured with bungee cords. (*Sergeant David 'Bob' Monkhouse, with kind permission RDG*)

This photo was taken on the last day of 2004, and it shows an RDG crew on patrol in Iraq. The turret has a wire cutter fitted to the smoke discharger bracket on the loader's side. The Challengers were used extensively for patrolling and the example seen here carries a thick coat of dust and rather battered front mudguards. (*Sergeant David 'Bob' Monkhouse, with kind permission RDG*)

The other very obvious change in the Challenger 2's appearance by 2004 was that the desert sand paint scheme used in the invasion was discontinued. Traces of the sand paint remained on the ROMOR reactive armour cradle and on the frames carrying the Chobham side armour (which came from stocks left from the 1991 Gulf War). This photo was taken in March 2005. (*P.D. Drabble, with kind permission RDG*)

The use of dust skirts on the hull sides ended soon after the original invasion. The Challengers seen here belonged to the Royal Dragoon Guards and were photographed undergoing maintenance in early 2005. *(P.D. Drabble, with kind permission RDG)*

A Royal Scots Dragoon Guards Challenger 2 undergoing maintenance in Iraq. We can see how the appliqué armour had to be removed to perform many maintenance tasks involving the suspension. (*Crown Copyright*)

'Streetfighter'. Several different Dorchester armour kit combinations can be fitted to the Challenger 2, and the example seen here has the front Chobham armour pack fitted. Bar armour is carried on the turret and hull rear but we can see the remote weapons station, IED jammers, bird table and cameras fitted. (*Tim Neate*)

The bird table is a large antenna that functions as part of the Challenger 2's urban warfare equipment. The rear of the vehicle is protected against RPG attack by simple and effective bar armour. Auxiliary fuel drums are not generally fitted for urban combat, for obvious reasons. (*Tim Neate*)

The ultimate solution to the lower front plate's lack of Chobham armour was the addition of a massive appliqué pack that could be fitted to any Challenger 2 as part of the Dorchester 2H armour kits that are presently employed to bring the vehicle to Theatre Entry Standard (TES). (*Tim Neate*)

As the Challenger 2 received its various armour upgrades in Iraq the vehicle's weight increased substantially from its factory weight of 62 tons. The severe damage inflicted by IED attacks in 2006 resulted in the Challenger 2 receiving a substantial armour upgrade from the layout seen here. The CV12 engine, the transmission and the suspension received close attention for signs of wear but in general the rate of component wear did not increase noticeably. The armour upgrades in use by 2007 pushed the Challenger 2's weight beyond 70 tons. (*P.D. Drabble, with kind permission RDG*)

The Royal Scots Dragoon Guards on ranges near Basra, 17 November 2008. After Telic 1 (as the 2003 invasion became known retrospectively), the Challenger force in Iraq dropped to a single squadron. The tanks could be deployed wherever they were required in the British occupation zone and they saw extensive use. (*U.S. Army photo by Sergeant Gustavo Olgiati*)

A Royal Scots Dragoon Guards Challenger II fires its main gun on a target on 17 November 2008, in Basra, Iraq. (*U.S. Army photo by Sergeant Gustavo Olgiati*)

The turret sides of the Challenger 2 seen here in Iraq in 2008 have been strengthened against the threat of RPG attack with bolt-on armour blocks. The armour has been upgraded with a new Chobham pack on the hull nose, and the hull side appliqué protection has been substantially reinforced to protect from tandem warhead hollow charge attack. The use of electronic countermeasures and remote vision cameras by this time improved the tank's protection in an urban warfare environment at least as substantially as the advanced armour. (*U.S. Army photo by Sergeant Gustavo Olgiati*)

A Royal Scots Dragoon Guards Challenger II Main Battle Tank, moves across the ranges on 17 November 2008, in Basra, Iraq. (*U.S. Army photo by Sergeant Gustavo Olgiati*)

A Challenger 2 firing on the ranges in Iraq. (*U.S. Army photo by Sergeant Gustavo Olgiati*)

The muzzle flash and smoke from the muzzle. This tank carried the full armour kit developed through years of urban operations around Basra. (U.S. Army photo by Sergeant Gustavo Olgiati)

This photo was also taken by a United States Army photographer on 18 November 2008 near Basra, showing a SCOTS DG Challenger 2 during a range period. The extra armour does result in faster wear on tracks and on roadwheel tyres. (U.S. Army photo by Sergeant Gustavo Olgiati)

This is the desert version of the UK tank suit, virtually the same design specified for temperate climates but in sand colour, and made of slightly lighter material. *(Robert Griffin courtesy Home Headquarters RDG)*

A Royal Dragoon Guardsman wearing the helmet and body armour typical for Operation Telic in desert DPM pattern. The driver and loader normally carry SA80 series rifles or L22A2 carbines, while the gunner and commander are normally armed with the 9mm pistol. Since the late 1940s the RAC was issued the Browning GP35 based L9 series, but these have been recently replaced with the Glock 17 Generation 4 pistol made in Austria. It is conceivable that the crew could dismount the loader's GPMG in any emergency requiring the abandonment of their tank. Because the existing stocks of body armour and uniform items will be worn for some time to come, many photos of crewmen on exercise taken in the 2011-2015 period show a mixture of camouflage patterns. (*Robert Griffin courtesy Home Headquarters RDG*)

Crewmen wearing the DPM one piece tank overall (training version). The nomex tank overalls worn on operations are very expensive and usually are only issued for deployments into combat zones. They also are wearing the AFV crew vest, which is similar to other assault vests but has a wraparound green outer cover to make it easier for the crew to enter and exit the tank. (*Crown Copyright MoD*)

# Chapter Fourteen

# Challenger 2 Variants

The only two operational variants of the Challenger 2 ordered for the British Army are the Trojan AVRE and the Titan AVLB both of which are used by the Royal Engineers. Thirty-three Titans and thirty-three Trojans were delivered by 2006 to replace the old Chieftain based bridge layer and the Chieftain Armoured Vehicle Royal Engineers (AVRE). The improvements in mobility and capability that these offer the sappers are significant and the Trojan has done sterling work in Afghanistan. Normally these vehicles are attached to armoured battle groups for obstacle crossing or dealing with minefields.[72]

Two other variants were built; one to conform to customer requirements (*Project Copenhagen*) and the second (*Project Exmouth*) to experiment with a different power pack and cooling system. The only export version of the Challenger 2, despite many sales attempts, was the Challenger 2 with alternate cooling for extreme desert conditions. *Project Copenhagen* was conducted by modifying the V1 prototype to meet the requirements of the Royal Omani Army. One issue that had to be addressed was the ability to maintain the full 1,200HP output of the CV-12 in the heat and dust of the Omani desert. To this end the Omani vehicle had a much revised cooling system and rear hull grills shared with the Challenger 2E. Another difference to the British Challenger 2 configuration was the Royal Omani Army's request for the retention of the single pin track from Challenger 1. This resulted from the CRARRV recovery vehicle's tremendous resistance to track shedding during the Omani Challenger 2 trials with Prototype V9 in 1992. The standard double pad tracks used in the trials proved far easier to shed in Omani terrain than the old pattern carried by the CRARRV; accordingly the test crews and support personnel were instructed to fit V9 with CRARRV sprockets and tracks. These became standard for the Omani production order. The Omani Challenger 2s also carry the old style bazooka plates that were standard on the Challenger 1 and the loader's machine-gun mounting is fitted for the .50-Calibre M2 Browning HMG.

The second gun tank variant was known as Challenger 2E (for *Europack*) and was the final version of the gun tank offered to foreign customers by BAe. In essence Vickers took the Prototype V9 definition vehicle after its Middle Eastern marketing tour in 1992 and replaced the Perkins CV-12 power pack with the 1500BHP MTU 883 V-12 diesel from the Leopard 2. The German diesel was coupled with a transversely mounted Renk HSWL 295 automatic transmission and a new cooling system. The new engine required a substantial redesign of the hull to incorporate large louvres and it totally changed the appearance of the rear of the tank. Stowage items from the rear plate were relocated elsewhere on the hull; the arrangement also meant that any auxiliary fuel drums needed fitting to the top of the hull rear rather than in the usual lower location – a method of carriage already employed on the tanks built for the Royal Omani Army.[73]

Different sights have been proposed for the Challenger 2E's gunner (SAVAN 15) and commander (VS580-10) both of which feature thermal channels and can be used with or without the TOGS 2 system. Vickers has always been at pains to point out that Challenger 2E was a concept vehicle and V9 was modified over a number of years before being employed as the British entry for the Hellenic Army MBT Trial. It served its purpose and much data on operating vehicles in hot climates was obtained. When parked running next to a conventional Challenger 2 its quietness was particularly noticeable.

In 1999 Vickers Defence was bought by Rolls-Royce and was sold on to Alvis in 2002 as Alvis-Vickers. In 2004 Alvis-Vickers was taken over by BAe Systems and was grouped between 2004 and 2005 with Royal Ordnance, Alvis-Vickers and the American firm United Defence to operate as BAe Land and Armament Systems Group. BAe now owned the Challenger 2 and its future. Attempts to sell the Challenger 2 to other armies have been disappointing as significant NATO reductions resulted in large quantities of low-priced Leopard 2 and M1A1 tanks becoming available at the same time.[74]

Between August and October 1992 the Challenger was marketed in Kuwait, Saudi Arabia, Oman and the United Arab Emirates, with a series of demonstrations using Prototype V9 culminating in the successful Omani order. Later efforts to sell the tank focused on an increasingly slender market despite generous financing and co-production offers supported by the government. While Alvis-Vickers and other contractors made every effort they always faced stiff competition from Krauss-Maffei's reconditioned Leopard 2s. In Project *Aorta*, the South African Defence Force's selection process for a new main battle tank, the Challenger 2E seemed the likely successful contender before the entire programme was suspended after government changes. The Hellenic armed forces were looking to replace their ageing M48 based fleet and a competitive evaluation trial was arranged in 2003. Designs from most tank producing nations were represented; these included the M1A2 Abrams, the Leopard 2A5, Leclerc EAU, T-80UE, T-84 and the Challenger 2E (based on the V9 prototype hull with a modified production Challenger 2 turret fitted). Final deliveries of Challenger 2s for the British order had been made in 2002 but the production line in Newcastle was still building the bridge-layer and engineers' versions of the Challenger 2; series production of the Challenger 2E gun tank would have followed had an order been secured.

Of the six MBT types evaluated during the trial, given a maximum possible operational and technical score of 100%, the best performers were: Leopard 2A5, 78.65%; M1A2 Abrams, 72.21%; Leclerc EAU, 72.03%; and Challenger 2E, 69.19%. Next came the T-84 and last the T-80UE. The Leopard 2A5 was the only one to demonstrate deep fording capability, while the M1A2 had the best firing results during hunter/killer target engagements. The German 1,500hp MTU Euro Power Pack was fitted in both the Leclerc EAU and the Challenger 2E, and these two vehicles had the best cruising range and lowest fuel consumption. The T-80U had the best mobility and reliability. Some shooting results for the trials are also known: Leopard 2A5 (80% hits), Leclerc: (65%), T-84(47%), Challenger 2E (40%), No data exists for the T-80UE and M1A2 but the question is how the Challenger 2E scored so poorly. It was fitted with upgraded sights and fire controls and the vehicle was in top form throughout the trial so the answer was frustratingly clear.

The Challenger 2E was not provided with standard L30A1 CHARM 3 ammunition for the trial. The Vickers team used left-over L11A5 120mm L26 APFSDS-T CHARM1 ammunition from Operation Granby gifted by the Ministry of Defence. This ammunition dated from 1991 but was fully compatible to the later gun (as with any L11 APFSDS ammunition). There were comparable ammunition problems with some of the other contestants. The T-80UE and T-84 only used practice 3P31 rounds, which only correspond to BM15 service rounds at ranges of less than 1500 metres. The M1A2 Abrams could not fire practice ammunition at all and waited for three days before live supplies were brought in. The whole trial was mired in controversy and was won by the Leopard 2A5. Allowing for the inevitable complaints from the losers it seems strange that the British did not make a better effort. There was no explanation for having brought their best product to a trial and then using old ammunition for an earlier gun. Much nonsense was postulated by armchair experts on military forums but it seems the ghosts of CAT87 were never so far away![75]

BAe ceased to offer the Challenger 2E in 2005. The hard truth is that Britain was poor at selling the Challenger 2 in a period when cheaper surplus weapons were widely available. It has also been suggested that the British government secretly sought failure in order to close the tank producing factories. The tank industry that produced the Challenger 2 has now shrunk massively; foreign ownership and mergers have caused the much diminished overall defence industry to concentrate on modernising existing armaments rather than building new weapons and the procurement of modified foreign designs is much more widespread than had previously been conceivable.[76]

A photograph of the lower glacis plate of V-9, then marketed as the 'Desert Defender' to Kuwait, showing the registration in Arabic in August 1992. (D. Lunn)

Front view of the Desert Defender in August 1992, showing the original style of loaders machine-gun mounting and the thermal covering over the forward area of the turret. (*D. Lunn*)

An unusual setting for a Challenger 2, this is V9 on its Middle Eastern sales trip with a CRARRV to its right. These vehicles were photographed outside the British Embassy in Kuwait 1992. (*D. Lunn*)

The Challenger 2E was the export version of the basic Challenger 2 proposed for foreign customers. Only one Challenger 2E was built from prototype V9, and the Hellenic Army inspired camouflage scheme seen here in the period prior to the Hellenic Army MBT trials may in fact be worn by the prototype configured to Omani standard, prototype V1. *(Peter Brown)*

One of the configurations proposed for the Challenger 2E turret in the late 1990s prior to the Greece trials is shown here. No TOGS system is fitted and the appearance of the turret recalls its Mk.7 origins. The hull could be V1s or V9s. *(J.W. DeBoer)*

The M2 series Browning 12.7mm heavy machine gun installed on the loader's hatch is notable. We can also see that the VS580-10 commander's panoramic sight is fitted along with the SAVAN 15 gunner's primary sight. The driver could presumably also be provided with a thermal viewer if required. (J.W. DeBoer)

A close-up of the VS580-10 sight and the SAVAN 15 gunnery sight. This configuration did not require the TOGS sight over the mantlet and it is possible we might see a similar configuration adopted for British Challenger 2s as the original electronics and fire controls are eventually modernised. (J.W. DeBoer)

A close-up of the mantlet without the TOGS barbette fitted. (J.W. DeBoer)

Presumably a production version of the Challenger 2E fitted with these sights would have provided armoured shutters to protect them from artillery splinters or small arms fire, and images taken in 1999 show these modifications in Vickers publicity literature. (J.W. DeBoer)

*Above:* Both the Omani production Challenger 2 and the Challenger 2E hulls incorporated a very different engine compartment layout to a standard British Army Challenger 2. The rear of the hull incorporates large cooling louvres, and carries the registration number 06SP87. (*J.W. DeBoer*)

*Right:* We can see the different arrangement used for the rear hull lights, mounted on a jerry can rack carried on each side of the rear plate, along with the first aid box. (*J.W. DeBoer*)

Auxiliary fuel drums are carried on the rear edge of the engine deck rather than on the hull rear proper. (J.W. DeBoer)

The auxiliary fuel drums were strapped into place in a similar manner to the drums carried on the rear of British Challenger 2s. The filled drums would have been lifted onto the rear deck and strapped into place. They could be pushed off when empty. (J.W. DeBoer)

*Right:* Side view of the left hand rear jerrycan rack and light cluster also shows the mounting brackets for the auxiliary fuel drums. (J.W. DeBoer)

*Below:* The mounted auxiliary fuel drums would have interfered with turret traverse over the rear arc of the vehicle and were intended to be discarded prior to combat. (J.W. DeBoer)

Despite proving many of the concepts later considered for British upgrades to the Challenger 2 and the provision of generous terms, the Challenger 2 came at the wrong time. Defence budgets shrank, the surplus MBT market was full of cheaper options, and Vickers itself was soon forced to merge with Alvis and eventually stopped producing MBTs. Note the retention of Challenger 1 style bazooka plates and the tow bar stowage. (*Lawrence Skuse*)

The Challenger 2E at the MVEE display in 1999. Note the absence of the TOGS system, the provision of the front dust skirts and the disruptive camouflage scheme. (*Lawrence Skuse*)

The CRRARV was also offered for export at the same time. Sales did not ensue. (*Lawrence Skuse*)

The Desert Defender (prototype V9) and its attendant CRARRV lined up on the firing point in Oman in 1992. Those who are familiar with tank firing range layouts will notice the striking similarity to a normal tank range in Europe. (*D Lunn*)

V9 posing on its run up to the firing point, ready to show off its gunnery capabilities. (D. Lunn)

Seen again during the lead-up to the firing demonstration, we can see the thermal cover on the turret and the meteorological probe in its raised position. (D. Lunn)

One method that was always sure to make a tank commander popular in BAOR was to do a neutral turn on a thinly laid tarmac road in hot weather. Despite the heat, in this case the Challenger 2 prototype has surprisingly not caused any damage to the road. (D. Lunn)

One prototype vehicle (V9 which was later modified into the unique Challenger 2E) was used for most of the sales trials. A major difference between this version and the one that Oman finally ordered was the Omani Challenger 2's use of single pin track rather than the double pin type used on the British MBT, and the use of a louvred hull rear plate to assist cooling. (D. Lunn)

A good view of what tank maintenance looks like: tool kits and rags lying by the road wheels – and a luxury seen no doubt for the sake of the Omani trials, a set of steps to save the technicians from clambering on and off the vehicle. (D. Lunn)

This is V9 again viewed from the other side; notice the small items that make the task of maintenance just that little more bearable. By the rear wheels we can see the ubiquitous army green canvas and metal folding chair, preferable by far to sitting on the hot tarmac. (D. Lunn)

This photo shows the Challenger 1 style out riggers that supported the bottom of the bazooka plates on the prototype vehicles, which were not present on the production Challenger 2 configuration ordered by the British Army, but which were ordered by the Omanis. (*D. Lunn*)

The CRARRV recovery vehicle and the V9 prototype Challenger 2 seen side by side in the desert in Oman in 1992. The CRARRV has turned out to be a most successful and powerful recovery vehicle despite being rushed into service for the first Gulf War in 1990. (*D. Lunn*)

With the gun over the right side we can see one of the turret stowage bins in the open position and also the location of the metrological probe. (D Lunn)

The thermal covers carried by V9 in the Middle East in 1992 were attached by Velcro type fastenings and were removable. It is interesting to note that the barracuda nets fitted to Challenger 2s today employ the same means of attachment. (D. Lunn)

Extensive maintenance was required throughout the trials and was carried out by contractors' technicians as well as by army personnel. This photo taken in Oman shows a scene any tank crewman will recognise, bazooka plates on the floor and rags lying all around the work area. There is one thing that tanks are not, and that is clean. (D. Lunn)

Maintenance underway in Oman. V9 was comprehensively transformed with a modified rear hull to suit Omani cooling requirements; upon its return to Britain it received further modifications for the Europack, becoming the Challenger 2E. (D. Lunn)

The original style of GPMG mount for the loader can be seen from this shot and also the Arabic script near the fire extinguisher handle. (*D. Lunn*)

A view seldom seen apart from by those who use the Challenger 2 (and especially by those who drive them) – the lower half of the heavily armoured mantlet, in this case on prototype V9 in Oman in late 1992. (*D. Lunn*)

The desert environment of Oman highlights the clean rear engine decking of Challenger 2 V9. (D. Lunn)

V9 at rest in 1992. We have a good view of how the gun clamp for the L30A1 gun secures it; also to the right can just be seen the muzzle cover which helps keep the bore clean when not actually firing. It is usually taken out for more tactical exercises, so the barrel "looks the part". (D. Lunn)

A scene inside a hangar during the Hellenic Army trial for its new MBT in 2003. The CRARRV is carrying out a pack change or engine maintenance on the Europack fitted to the Challenger 2E. (*D. Lunn*)

A group of officers taking a keen interest in the commander's station on the Challenger 2E during the Hellenic Army trials. Notice the thermal covering on the turret which is designed to reduce internal temperatures. The distinctive rubber side skirts and the rear hull louvres employed with the Europack are also visible. (*D. Lunn*)

A side view of the Challenger 2E while taking part in the Hellenic Army trials; note the fitting of the venerable Browning M2 .50 MG on the loader's station. (*D. Lunn*)

The universal comment from serving Challenger 2 crewmen on seeing the stowage layout of the Challenger 2E was jealously to point out the recovery bar and connector in its new sensible and accessible location on the rear of the hull side (instead of on the rear plate of a standard British Army Challenger 2, where the equipment was inevitably covered in mud by the time it was needed for recovery!). (*D. Lunn*)

A unique feature of the Challenger 2E is the provision of small louvres on the rear hull side required for the Europack. (*D. Lunn*)

It is hard to believe that this is Greece; the scene looks more like an arid desert. We can see that the thermal cover is also installed on the upper glacis as well as on the front half of the turret. (*D. Lunn*)

The blue sky and hot sun in this 2003 picture would have probably made the heat quite uncomfortable inside the MBTs being trialled. (*D. Lunn*)

Wearing a unique camouflage paint pattern, this Challenger 2 was photographed during the evaluation phase for Project Aorta, the South African MBT programme. The order was nearly made but South African politics put Project Aorta on hold in the first years of the new century. Should it be reactivated there is now no capacity to build new Challenger 2s. Leopard 2 or M1A1 will probably be adopted when the time comes. (*Private Collection*)

A Trojan being demonstrated at Gallows Hill in 2005. The hydraulic arm can lift vehicles like this with ease. (*Tim Neate*)

This shot shows the tie down arrangements for transporting the Trojan on a transporter and the different rear end and exhaust system used on the Trojan. (*R Griffin*)

A stock (therefore uncluttered) Trojan at what appears to be a military demonstration; it has its excavator bucket in the extended mode. (*Lawrence Skuse*)

The bare sides of this Trojan reveal the attachment points for additional equipment that could be fitted, including armour modules. (*Lawrence Skuse*)

Seen at Gore's Cross on the Salisbury Plain Training Area in 2008, this Trojan is pulling an all-terrain trailer. This allows the Trojan to carry additional fascines for gap crossing or any other supplies it might need. (*Tim Neate*)

The Trojan's chassis is that of the Challenger 2 MBT, but it differs in most other aspects of its hull construction. Because the upper surfaces of the hull are dedicated to the onboard storage of equipment, the rear hull plate incorporates the cooling louvres and air intakes. The vehicle's exhaust system remains on the sides of the engine compartment. (*Tim Neate*)

During a demonstration at Gallows Hill in 2005, this Trojan is shown conducting mine clearance and deploying a fascine simultaneously. The cleared lane is marked by high visibility markers deployed on either side of the vehicle's rear to the full width of the ploughed lane. (*Tim Neate*)

The Trojan can be fitted with supplementary protection to make it less vulnerable to infantry antitank weapons. The armour array demonstrated here is marked with triangular markings that may denote that they are not functional armour packs and that they are weighted strictly for demonstration purposes. (*Tim Neate*)

*Above:* A Trojan photographed at Perham Down in 2008. The massive size of the vehicle is immediately apparent. Because the basic superstructure is not armoured to resist more than mine blasts, artillery splinters and small arms, it has to be deployed with covering forces. The basic immunity level can be augmented substantially with Dorchester armour kits. (*Tim Neate*)

*Left:* A Titan AVLB deploying its bridge during a demonstration at Gallows Hill in 2006. The Titan replaced the Chieftain Armoured Vehicle Launched Bridge and is one of the most modern armoured bridge laying systems in service anywhere. (*Tim Neate*)

The Titan can launch the No 10, No 11 and No 12 bridges with equal ease and the largest of these can span a gap 26 metres wide. (*Tim Neate*)

The sides of the Titan are covered in mounting points for Dorchester armour kits. This vehicle was photographed at Perham Down in 2008. (*Tim Neate*)

The twin set of forward looking cameras mounted on the Titan's glacis; the introduction of rear and forward looking cameras has made the job of manoeuvring heavy armoured vehicles much easier when closed down. *(Lawrence Skuse)*

A close-up view of the front end of the Titan, notice the headlight installations are similar to that of the Trojan. The hydraulic bridge lifting gear on the Titan is visible; the long section on the ground with the plunger at the front is driven between the sections of the bridge, the cone engages in a hole in the bridge cross member, thus securing and aligning the bridge. All this can be done while the vehicle is closed down. The Titan employs the Br90 bridge system. Unlike previous Royal Engineers vehicles simply adapted from production battle tanks, the Titan and Trojan were specially designed and developed by BAe and feature a great deal of mission specific equipment. *(Lawrence Skuse)*

A well-worn section of the bridge launching assembly – once this heavily rusted part is on the ground the bridge will be in the vertical position, Titan then can slightly adjust its position to ensure a good base for launching its bridge. (*Lawrence Skuse*)

The Titan can be equipped with a front-mounted dozer blade for other engineering tasks that can be performed when not laden with its bridging equipment. (*Tim Neate*)

The Titan looks very empty when not laden with its bridge, and the vehicle layout is very simple as can be seen here. The Titan is very stoutly built as are all of its hydraulic components. It is expected to perform very demanding jobs and projected to remain in service for decades. (*Lawrence Skuse*)

This Titan AVLB is carrying two No 12 bridges, which can be launched in under 90 seconds. (*Published by the Ministry of Defence © Crown Copyright 2006. Reproduced under Open Government License*)

A Titan launching a No 10 scissors bridge. If required to cross a larger gap, a trestle can be placed in the gap and the bridge layed onto that, and another Titan can then drive onto the bridge and lay a further section. It is an extremely flexible system that gives the Royal Engineers a much expanded gap crossing capability. (Published by the Ministry of Defence © Crown Copyright 2006. Reproduced under Open Government Licence)

A Titan Armoured Vehicle Launched Bridge carrying a No. 10 bridge, driving in wet muddy conditions, showing just how much carrying the bridge restricts both the driver's and the vehicle commander's vision. (Published by the Ministry of Defence © Crown Copyright 2006. Reproduced under Open Government License)

Chapter Fifteen

# The Challenger 2's career since 2003

Much has changed in the British Army since the invasion of Iraq. The need for a MBT in the British Army has remained but that for large tank forces in NATO has been forgotten. While Operation Telic ran its long course a simultaneous campaign in Afghanistan (known in the British Army as Operation Herrick) drew a considerable percentage of the army's shrinking manpower and kept armoured regiments out of their roles for substantial periods. The Royal Armoured Corps underwent cuts in 2003 when The Royal Lancers were converted to armoured reconnaissance and seven MBT squadrons were cut from the remaining armoured regiments to reduce costs. The trend towards operating fewer MBTs was confirmed when the 4th Armoured Brigade became the 4th Mechanised Brigade in 2004. The army was reorganising itself into multirole brigades to deal with the strategic commitments of the new millennium on a tightening budget.

The RAC felt keenly the impact of the financial crisis of 2008. Ultimately this was revealed in the Strategic Defence and Security Review of 2010, which cut the operational MBT fleet to around 200 tanks and left most of the rest to be stored for use as spare parts. That year the Army 2020 plan proposed an 8-brigade army, and the maintenance of 3 divisional headquarters structures to allow for expeditionary and defence commitments. These plans were controversial and a parliamentary committee report warned that British forces were at considerable risk of being unable to meet the government's global commitments:

> We note that there is mounting concern that the UK Armed Forces may be falling below the minimum utility required to deliver the commitments that they are currently being tasked to carry out let alone the tasks they are likely to face between 2015 to 2020 when it is acknowledged that there will be capability gaps.[77]

The budget cuts affected all of the armed services and resulted in armoured units being used out of role for prolonged periods in Afghanistan. Naturally this affected the proficiency of MBT crews although simulators did provide some useful training aids. The use of RAC troops in so many roles was caused by the reduction of combat troops after the army cuts and the need to meet extensive commitments. The period since 2008 has been hard for the Challenger regiments. The total number of operational vehicles was still substantial but actual operation of these was kept to the minimum possible in order to save money and the vehicles themselves have received very little upgrading. Up to 40 Challenger 2s have been kept at BATUS and perhaps 60 remained in Germany prior to the complete return of British Forces. The presumption is that up to 160 operational vehicles remain in the UK.[78]

Today the British Army fields three Type 56 armoured regiments (each equipped with three squadrons of Challenger 2s) on paper but many of its armoured vehicles are held in storage and reserve. Each Type 56 Regiment formed the spearhead of an armoured infantry brigade in the 3rd (UK) Division to be the army's quick reaction force. The division was composed of three Strike Brigades. The 1st Armoured Infantry Brigade includes The Royal Tank Regiment, the 12th Armoured Infantry Brigade includes The King's Royal Hussars, and the 20th Armoured Infantry Brigade includes The Queen's Royal Hussars. The other Challenger 2 regiments established in the late 1990s have either been amalgamated or have slowly been re-equipped with light AFVs since 2005. The 1st (UK) Armoured Division has reverted to a non-armoured division.[79]

The training year in the Type 56 Regiments starts with range periods and then culminates in battle group training in Canada or possibly in Poland. The regiments have operated at much less than full strength for long periods and reservists have served sometimes as unpaid volunteers. In 2012 the governments' chilling future requirement for perhaps no more than 135 operational Challenger 2s resulted in dozens of tanks being mothballed or stripped for spares. Early Challenger 2s were scrapped and it is believed that less than 260 British Army Challenger 2s are maintained in operational condition. This follows the same theme of fleet reduction seen in France and Germany.[80]

This landing exercise in October 2010 was organized for an Immediate Staff Command Course. (Published by the Ministry of Defence © Crown Copyright 2010. Reproduced under Open Government Licence)

A Challenger 2 of the King's Royal Hussars climbing aboard a tank transporter at Tidworth in 2006. (*Published by the Ministry of Defence © Crown Copyright 2006. Reproduced under Open Government Licence*)

Seen here during an urban combat exercise conducted jointly with the French Army at the Sissonne CENZUB urban combat training centre in France, a Challenger 2 equipped with a Pearson dozer blade lays down smoke as it pushes through a barricade. (*Published by the Ministry of Defence © Crown Copyright 2012. Reproduced under Open Government Licence*)

With infantry positions in the foreground, this 1st RTR Challenger 2 named *Apolyon* was seen on exercise on Salisbury Plain in 2012. The amber flasher on the turret roof is a safety feature commonly carried for road moves. (*Published by the Ministry of Defence © Crown Copyright 2012. Reproduced under Open Government Licence*)

A Trojan AVRE on operations in 2010 in Afghanistan. The mine plough is fitted with a massive pipe fascine carried on its back. This could be deployed with the hydraulic arm when required. (*Published by the Ministry of Defence © Crown Copyright 2010. Reproduced under Open Government Licence*)

The Trojan AVRE proved well up to its tasks in Afghanistan. Notice the full width mine plough, the fascine and the Chobham armour appliqué packs and bar armour. (*Published by the Ministry of Defence © Crown Copyright 2010. Reproduced under Open Government Licence*)

The cloud of dust seen here is an indication of the sort of conditions that an AVRE might encounter; notice on the vehicle top a variation of the remote weapon station is fitted for local defence, and the mine plough is carried in the raised travel position. (*Published by the Ministry of Defence © Crown Copyright 2010. Reproduced under Open Government Licence*)

The mine plough carried by the Trojan is a massive piece of equipment and seen here in Afghanistan in 2013. (*Published by the Ministry of Defence © Crown Copyright 2013. Reproduced under Open Government Licence*)

A Trojan photographed during Exercise Lion Strike in 2014 on Salisbury Plain. It does not carry the heavy appliqué armour kits seen on operations, which reduces wear on the suspension. (*Published by the Ministry of Defence © Crown Copyright 2014. Reproduced under Open Government Licence*)

A Challenger 2 main battle tank with a Polish Leopard 2A4 tank on rough terrain in Poland during *Exercise Black Eagle* in 2015. Some 1,300 UK soldiers took part in the largest armoured deployment to Eastern Europe since 2008. The King's Royal Hussars Battlegroup served under the command of the 10th Polish Armoured Cavalry Brigade and alongside the 1st Polish Tank Battalion at the Zagan Training area in south-west Poland. (*Published by the Ministry of Defence © Crown Copyright 2014. Reproduced under Open Government Licence*)

Challenger 2 main battle tank moving quickly over rough terrain in Poland during *Exercise Black Eagle*. (*Published by the Ministry of Defence © Crown Copyright 2014. Reproduced under Open Government Licence*)

British Army Challenger 2 main battle tanks form up on the Polish landscape at the start of *Exercise Black Eagle*. This included both dry-training and live-firing and was designed to develop interoperability between the two Armed Forces. (*Published by the Ministry of Defence © Crown Copyright 2014. Reproduced under Open Government Licence*)

A maintenance stop during exercises in Poland. (*Royal Dragoon Guards*)

While the circumstances of this 'attack' during training in Poland are unknown, we can be sure that the Challenger 2's armour was not tested. The crew's patience may not have held out as well. (*Royal Dragoon Guards*)

A Challenger 2 at BATUS (British Army Training Unit Suffield), complete with fire simulation equipment. The British Army's base at Suffield, Alberta in central Canada is a massive expanse of firing ranges and exercise areas that have served to develop and refine British battle group tactics since the early 1970s. (*Published by the Ministry of Defence © Crown Copyright 2012. Reproduced under Open Government Licence*)

The Challenger 2 can be utilised to its maximum training potential at BATUS, and live ammunition is a regular part of the realistic exercise regime followed by each visiting battlegroup. Battlegroup structure can be assembled around the headquarters of an armoured squadron or a mechanized infantry company – allowing armoured and mechanised troops to train with each other for weeks. (*Published by the Ministry of Defence © Crown Copyright 2004. Reproduced under Open Government Licence*)

Photographed during a break in a training exercise in 2008, a MAN fuel tanker truck refuels a Challenger 2 Main Battle Tank at British Army Training Unit Suffield. (*Published by the Ministry of Defence © Crown Copyright 2008. Reproduced under Open Government Licence*)

The combined might of the 1 Yorkshire Regiment Battle group on display during Ex Prairie Lightning in 2014 at BATUS. Amongst the vehicles in view are Challenger 2 MBTs, Scimitar CVRTs, AS-90 155mm Self-Propelled Guns, Warriors, and even two MLRS launchers. Note how the last skirting plate is mounted upside down, a common practice at BATUS in recent years. (*Published by the Ministry of Defence © Crown Copyright 2014. Reproduced under Open Government Licence*)

Since 2011 the Challenger 2 fleet at BATUS has adopted an overall sand paint scheme. While vehicle names at BATUS seem more the exception than the rule, this lead vehicle carries the name *Danger Mouse* (presumably after the cartoon series). (*Published by the Ministry of Defence © Crown Copyright 2011. Reproduced under Open Government Licence*)

The British Army's Canadian training ground offers some unique scenery. The Aurora Borealis makes for a unique backdrop to this well-worn Challenger 2. (*James Patterson*)

Men of the Royal Tank Regiment at BATUS in 2015. We can see that the uniform camouflage has changed to the new multicam camouflage pattern from the older DPM, and that the Challenger 2 now wears a uniform sand paint scheme. The use of stripes on the gun barrel to denote troop and squadron and of large call sign and zap code markings remains unchanged. (*Captain Matt Noone, RTR*)

The amalgamation ceremony of the 1st and 2nd Royal Tank regiments. We can see the stark contrast of *Nomad* next to the more usual black and green camouflage worn by the squadron command tanks. The black and green (or green and sand) schemes worn by the Challenger 2s of the Royal Tank Regiment do not follow a standard pattern and great variation in the camouflage can be seen within the regimental tank park. Since 1957 the Royal Tank Regiment has shrunk from eight battalions to a single battalion. (*Marilyn Suckling Gear*)

Rear view of a Royal Tank Regiment Challenger 2 taken in 2012. The Challenger 2's hull layout, unlike its turret, shows its Shir 2 lineage. Its rear plate incorporates a towing point flanked by bollard hooks for both transport and towing. Two track links are stowed below the towing bar. The fuel pump for the auxiliary fuel drums, the first aid box and the indicator lamps complete the equipment carried on this end of the tank. The hull (pannier to pannier) is 3.52 metres wide with standard skirts. (*Courtesy Royal Tank Regiment Crown Copyright MoD*)

*Above:* A tank from A Squadron 1st RTR seen in 2012 while serving at the OPFOR unit at Warminster. This tank is missing a skirt and has original style road wheels fitted to the first and last wheel stations. The mounting points for turret side appliqué kits have been added to permit the quick addition of heavy Chobham armour. (*Courtesy Royal Tank Regiment Crown Copyright MoD*)

*Right:* The Royal Scots Dragoon Guards held a welcome home parade down the Royal Mile on 4 July 2009. This excellent view of one of the participating Challenger 2s in Edinburgh was captured early that morning before the parade began. The uncluttered upper surfaces of the turret are clearly visible. Today the SCOTS DG are equipped with lighter AFVs as a result of the 2010 defence review. (*Adarsh Ramamurthy*)

The snowy ranges at Hohne in the winter in 2013 set the scene for this firing period for a Challenger 2 of the QRH. (*Published by the Ministry of Defence © Crown Copyright 2013. Reproduced under Open Government Licence*)

A Challenger 2 of the Queen's Royal Hussars photographed on the ranges in Germany in 2013. The QRH are the armoured regiment assigned to the 20th Armoured Infantry Brigade. (*Published by the Ministry of Defence © Crown Copyright 2013. Reproduced under Open Government Licence*)

The 20th Armoured Infantry Brigade is the only element of the RAC still based in Germany, and it is expected to return to the UK in the near future. (*Published by the Ministry of Defence © Crown Copyright 2013. Reproduced under Open Government Licence*)

The Royal Dragoon Guards converted to lighter armour as a result of the defence cuts outlined in 2010. The RDG Challenger 2 seen here in 2011 has the attachment points fitted to the turret sides permitting installation of Chobham armour appliqué packs. (*Tim Neate*)

The muzzle blast as a Challenger 2 fires its L30A1 120mm gun. (*Published by the Ministry of Defence © Crown Copyright 2014. Reproduced under Open Government Licence*)

A HESH round is fired by a Royal Tank Regiment Challenger 2 in 2014. This tank has the mounting points for the turret side appliqué armour installation. (*Published by the Ministry of Defence © Crown Copyright 2014. Reproduced under Open Government Licence*)

A Challenger 2 of Badger Squadron 2nd Royal Tank Regiment training on the Salisbury Plain Training Area. The badger's face device worn on the TOGS barbette was carried by this squadron, although sometimes it was carried in different locations on the tank. (*Published by the Ministry of Defence © Crown Copyright 2016. Reproduced under Open Government Licence*)

The Kings' Royal Hussars photographed during training with mechanised infantry in 2014. This Challenger 2 is making a smoke screen by injecting diesel oil into its exhaust during *Exercise Lion Strike*. (*Published by the Ministry of Defence © Crown Copyright 2014. Reproduced under Open Government Licence*)

An armoured column from Cyclops Squadron, the Royal Tank Regiment, deploying on a training exercise in 2016. (*Tim Neate*)

A Challenger 2 crewed by men of the Royal Wessex Yeomanry pauses to watch an RAF Chinook lift in underslung Wolf Scout Land Rovers on Salisbury Plain in 2014. (*Copyright UK Ministry of Defence. Published under Open Government License, 2014*)

# Chapter Sixteen

# The Future

Much of the discussion about the Challenger 2's future surrounds its gun, its electronics and its powertrain. The British Army expect to keep the Challenger 2 in service for at least fifteen and possibly twenty-five more years. It is well armoured and can be fitted with yet more effective armour types as required. It has proven itself as a suitable vehicle for the alternative Europack power plant. The potential gun and electronic upgrades that could be implemented are harder to define.

The L30A1 was Britain's final venture into an indigenous 120mm rifled gun, but this weapon is still extremely effective and will arm the Challenger 2 for years to come. Alternative armaments have been examined in the interest of cost reduction and NATO standardization. The 2005 Challenger Lethality Improvement Programme (CLIP) was undertaken to investigate the eventual replacement of the current L30A1 with the 120mm Rheinmetall L55 smoothbore gun used in the Leopard 2A6. The L55 fires rounds that have similar armour piercing capabilities to the L27A1 CHARM 3 round but the Germans have avoided depleted uranium in favour of long rod tungsten penetrators. New ammunition technology developed in 120mm smoothbore calibre in Israel, France and the United States offer a wide range of capabilities to MBTs armed with this weapon. A smoothbore gun would allow the Challenger 2 to use NATO standard ammunition developed in Germany and in the US without the negative environmental implications of depleted uranium rounds.[81]

A single Challenger 2 was fitted with the German L55 smoothbore and began trials with the ADTU at Bovington in January 2006. It is the same length as the L30A1 and fitted with the rifled gun's cradle, thermal sleeve, bore evacuator and muzzle reference system. Trials revealed that the German tungsten sabot DM53 round had a slightly better armour piercing performance round than the depleted uranium CHARM 3 round but that ammunition stowage was decreased to an unacceptable level without serious modification to the vehicle interior. The main advantage of the German 120mm gun was the longer term availability of ammunition types. Vast improvements in the range of smoothbore 120mm ammunition of all types had been made since the early days of the L44 Rheinmetall gun in the 1980s. In comparison, the L30A1's ammunition is restricted to three types developed over twenty years ago. While stocks of these suffice at present the Challenger 2's gun will presumably be re-examined.

In 2006, a figure of £386 million was estimated to refit all of the British Challenger 2s with the Rheinmetall gun. It was found that redesign of the Challenger 2 turret to accept the new ammunition was not practical on the existing budget although the weapon clearly worked. The MoD has quietly dropped the CLIP programme in the name of cost savings. If the Challenger 2 is ever to mount the German smoothbore gun as standard the internal arrangements of the hull and turret will need substantial change. The eventual adoption of the 120mm NATO gun carries the more likely implication of fitting a new turret to the Challenger 2. With the economic cuts the RAC is facing and the reduction of the army the Challenger 2 will probably continue as at present. As production lines for rifled 120mm ammunition in the UK were closed after BAe ceased to offer the Challenger 2E existing stocks of ammunition for the L30A1, while large, are finite. Thus the decision to cease producing rifled 120mm ammunition makes the rearmament of the Challenger 2 inevitable.[82]

The arrival of the T14 Armata in the Russian arsenal has forced a reconsideration of the short-sighted wishful thinking that has dominated the Army's (and NATO's) approach to the MBT in the past decade. The British government has recently reversed its neglectful stand on the main battle tank and will undertake an extensive upgrade of the Challenger 2's electronics, engine and protection over the next few years. The Challenger Life Extension Programme (CLEP) will extend to most of the tank park currently under the fleet maintenance programme which essentially functions as a centralised inventory of Challenger 2s that are drawn as required by the MBT regiments. This is similar to the mid-life upgrades planned for the Leclerc, M1A2 and for some of the armies employing the Leopard 2. The Challenger 2 may receive a new turret or the substantial modifications necessary to refit the British tank park with the 120mm smoothbore or possibly a new auto-loaded 130mm gun turret. Time will tell what modifications will be made to the Challenger 2's gun, fire controls, gun control system, optics, armour and powertrain but it is expected to serve until 2035 by when it may have evolved beyond recognition.[83]

The Challenger 2 was ordered to defeat any enemy tank it might meet in a conventional battle on the North German Plain. Sadly it was produced and came into service when the day of huge conventional armies in Europe was drawing to a close. As a result, it was never ordered in large numbers by the British Army and secured only a single order from abroad. The main battle tank still offers the soldiers of tomorrow a critically important weapon for the future battlefield – as it has for the past century. The Challenger 2 (like the Leclerc and to a lesser extent the more widely adopted Leopard 2 and M1 Abrams) will eventually be measured as a Cold War era design with finite development potential – as will all of its contemporaries. It remains the best tank Britain has ever produced and still amongst the best MBTs in service. It is a battle-proven and durable design which has the confidence of its crews. Its evolution is a matter of speculation but, regardless of this, the Challenger 2 will continue to serve the Royal Armoured Corps for years to come.

Theatre Entry Standard (TES) is the term used to describe the Challenger 2 with full appliqué armour suite and anti-IED equipment fitted. This most recent version of the heavy combination of reactive, passive and bar armour provides excellent protection from most battlefield threats. (*Published by the Ministry of Defence © Crown Copyright 2017. Reproduced under Open Government Licence*)

The Armoured Trials and Demonstration Unit based at Bovington maintains this Challenger 2 TES as a definition vehicle. This same standard of armour can be applied to Challenger 2s deploying on operations. The years to come may see the Challenger 2 substantially modified to meet the next generation of Russian MBTs. (*Published by the Ministry of Defence © Crown Copyright 2017. Reproduced under Open Government Licence*)

The TES armour is hidden beneath the new Mobile Camouflage System (MCS) employed to break up the Challenger 2's silhouette and enable its concealment on the battlefield. The great width of the vehicle with the full TES armour arrays is apparent. (*Published by the Ministry of Defence © Crown Copyright 2017. Reproduced under Open Government Licence*)

Even in the era of passive night vision, composite armour and depleted uranium long rod penetrator ammunition, the use of terrain and tactical skill is what wins battles. The TES equipment fitted to the Challenger 2 adds up to ten tons to its basic weight, which the CV12 can cope with without problems, even in the roughest terrain. (*Published by the Ministry of Defence © Crown Copyright 2017. Reproduced under Open Government Licence*)

The curved trajectory followed by the low velocity 120mm High Explosive Squash Head (or HESH) round fired from the L30A1's rifled barrel permits the Challenger 2 to engage enemy MBTs at great range. This TES Challenger 2 is making use of terrain to hide its bulk in a hull down position. The gun is at high elevation, very likely in anticipation of firing off a HESH round. We can see the bar armour deployed around the rear of the hull. (*Published by the Ministry of Defence © Crown Copyright 2017. Reproduced under Open Government Licence*)

The TES' anti IED equipment adds considerably to the height of the Challenger 2's silhouette. (*Published by the Ministry of Defence © Crown Copyright 2017. Reproduced under Open Government Licence*)

Bar armour protects the rear of the vehicle's hull and turret from infantry RPG type hollow charge weapons. Bar armour is cheap, relatively light and very effective. (*Published by the Ministry of Defence © Crown Copyright 2017. Reproduced under Open Government Licence*)

A more familiar, lighter armour standard is normally seen on Challenger 2s employed on exercise and in training. The TES standard armour package certainly increases mechanical wear and it is only issued as required for operations. This Challenger 2 was photographed in Exercise Venerable Gauntlet. Some 3000 personnel from 14 NATO countries were involved in the exercise validating NATO's Very High Readiness Joint Task Force (Land) expected to be activated in 2017. (*Published by the Ministry of Defence © Crown Copyright 2017. Reproduced under Open Government Licence*)

A full Challenger 2 squadron can be seen on the right hand side of this photo taken from a helicopter during the preparations for Exercise Tractable 2016. (*Published by the Ministry of Defence © Crown Copyright 2016. Reproduced under Open Government Licence*)

With bivvies erected on their engine decks and a variety of different camouflage schemes in evidence (typical of the modern 'issue as needed' MBT fleet) the crews of the Challenger 2s seen here are waiting for Exercise Tractable 2016 to start. (*Published by the Ministry of Defence © Crown Copyright 2016. Reproduced under Open Government Licence*)

Challenger 2s of the Royal Tank Regiment during the exercise. The barrel stripes are for tactical marking within the Royal Tank Regiment to indicate troop identity. (*Published by the Ministry of Defence © Crown Copyright 2016. Reproduced under Open Government Licence*)

The CRARRV can also be protected with appliqué kits. This ADTU CRARRV was displayed at Larkhill in 2015 with passive and bar type armour kits fitted. (*Tim Neate*)

The rear of the CRARRV, due to its mode of operation as a recovery vehicle, cannot be protected as extensively as the front and sides of the vehicle. (*Tim Neate*)

Theatre Entry Standard (TES) is a variable armour equipment standard that would be determined by the nature of the expected threat on the battlefield. This Challenger 2 is fitted with Dorchester 2H arrays on the turret and hull sides, as well as the remote weapon system – but is not fitted with the front appliqué Chobham block. (*Tim Neate*)

The rear of this Challenger 2 is also not equipped with bar armour and it does not carry the 'bird table' and anti-IED equipment. The Dorchester 2H armour on the hull and turret sides will still take the vehicle weight up towards 70 tons. (*Tim Neate*)

*Megatron* carries the most modern TES heavy armour suite, seen here at the Bovington Tankfest in 2013, and worth comparing to the armour carried in 2007 in Basra. The flank protection is placed to protect the fighting compartment and it now incorporates Dorchester 2H armour, supplemented by explosive reactive armour and bar armour to give maximum protection against kinetic energy attack, antitank guided missiles and RPG type weapons. (*Colin Rosenwould*)

Along with the extra armour, *Megatron* carries the 'Barracuda' type fixed camouflage system. Note the wire cutter fixed ahead of the commander's position. The armour protection that has proven a major component in the Challenger 2's success in combat is always evolving. (*Colin Rosenwould*)

We can see the massive Chobham block fitted to the lower front plate, as well as the velcro fixtures used to keep the Barracuda system fixed in place. The driver's video camera is protected by an armoured cover, and also visible are the headlamp covers to ensure that the driver's headlamps can be protected from damage in an urban combat environment. (*Colin Rosenwould*)

The Challenger 2's remote weapon system can be fitted with a 7.62mm NATO general purpose machine gun (as seen here), a .50 Calibre M2 Browning heavy machine gun, or a 40mm automatic grenade launcher. It replaces the loader's periscope and includes a full thermal vision capability, as well as the facility of being fired from under armour. (*Colin Rosenwould*)

The bar armour protection that is used on deployment to protect the rear of the hull sides is mounted with heavy brackets to the armoured skirt plates. (*Colin Rosenwould*)

The 'Bird Table' antennae system seen from beneath. Communications and electronics are of greater importance on the digital battlefield than ever. It is foreseeable that the Challenger 2 will eventually receive other electronic countermeasures, possibly with the goal of decoying or destroying incoming antitank guided missiles. Systems like the Israeli 'Trophy' have made significant advances in countering the latest generation of ATGMs. (*Colin Rosenwould*)

Because of the risk of fire, limitation of turret traverse and the configuration of the rear bar armour, the auxiliary fuel drums are not normally carried in the TES configuration. (*Colin Rosenwould*)

The most extensive version of the Dorchester 2H side armour, which incorporates Chobham type protection as well as reactive armour. It is laid out in a series of layers to give maximum protection to the crew fighting compartment and ammunition stowage, against both kinetic energy rounds and modern ATGMs. *Megatron*, as a demonstrator, is carrying inert armour arrays. (*Colin Rosenwould*)

The same approach is employed to protect the turret sides, which have been a traditionally vulnerable area on all tanks for RPG or ATGM strikes. Reactive armour and layered armour solutions have been effective in defeating hollow charge weapons. (*Colin Rosenwould*)

A side view of the right side turret appliqué armour mounting. The Perkins 1200HP CV12 engine is adequate for weight approaching 120% of the Challenger 2's original specification, a testament to its sound design. One possible easy upgrade to Challenger 2 would be to adopt the Leopard 2's 1500 HP diesel; this has already been proven and tested and gives a further reserve of power to extend the possible range of armour protection options. (*Colin Rosenwould*)

*Caen* of Cyclops Squadron, The Royal Tank Regiment, photographed during exercises on Salisbury Plain in 2016. (*Tim Neate*)

A King's Royal Hussars Challenger 2 in the summer of 2014. Note how one of the rear auxiliary fuel drums has been turned into a stowage bin, a typical peacetime convenience that allows the crew to carry a few creature comforts. (*Tim Neate*)

A Challenger 2 DTT being put through its paces. (*Tim Neate*)

# Notes

1. The Challenger 2's design concept reached a stage where prototype production could be considered just as the Cold War ended unexpectedly.
2. The British shared the Chobham armour technology with the USA and with the Federal Republic of Germany in the early 1970s. Ironically the American M1 was the first production tank to feature this revolutionary armour and Britain did not field a Chobham armoured tank until the Challenger 1 entered service over two years later. See National Archives File DEFE 24/1369. Vehicles. Chieftain replacement and its Armament. 26.09.78–31.10.78. The armament considered for the MBT80 evolved from the types explored in the earlier FMBT programme. One option was to have employed a 110mm rifled gun firing fixed ammunition although this was discarded in the mid-1970s with the promise of further development of the 120mm gun firing two piece ammunition. By September 1978 GSR 3752 which was intended to set the MBT 80's main armament and other characteristics was not finalized but the existing files from late 1978 indicate the intention to use a new rifled 120mm gun. This file notes MBT80 project definition stage to run 1979–1980, developmental stage 1980–1987, and series production start in 1989.
3. See National Archives File DEFE 24/1369. Vehicles. *Chieftain replacement and its Armament.* 26.09.78–31.10.78. This file notes that two different powerpack options were under consideration in mid-1978. The MBT80 hull would have very likely resembled the Shir 2's, although possibly with a slightly longer hull had a larger engine been developed. A major cause for delay was the lack of an existing engine more powerful than the CV12, and a second was that production of the FV 4030 for export would have certainly impacted capacity to produce the MBT80 prototypes! The MBT80 turret would have incorporated a panoramic commander's sight, hunter/killer fire controls and thermal sights for the commander and gunner.
4. For details of what the MBT80's prospective fire controls might have included see Griffin, Robert. Challenger 1 Volume 1. Kagero Publishing, Poland. 2013. See National Archives File DEFE 24/1369. Vehicles. Chieftain replacement and its Armament. 26.09.78–31.10.78. The MBT80's development was planned to be tested and proven by constructing fifteen prototypes built in two batches, the first of which was expected in 1983.
5. The FV4211 (or "Aluminium Chieftain") of 1970 predated the FMBT programme, and was in many respects the template that was followed by both the *Shir 2 and* the MBT80 concept – at least in so far as the general hull layout and armour arrangements were concerned. When it was constructed in 1969–1970 the FV4211 was one of the most advanced MBT designs and its use of an aluminium armour plate construction was both novel and the cause for its failure due to stress cracking. Fire controls were not given the government investment that was lavished on armour development but the private sector pursued improvement vigorously. Barr and Stroud and GEC Marconi were both active in the late 1970s improving optics and fire controls but, because of the internal dynamics of the government's arms procurement practices and their cost, none were adapted to the Chieftain or the *Shir 2* as they transformed into the Challenger.
6. See National Archives File DEFE 24/1369. Vehicles. Chieftain replacement and its Armament. 26.09.78–31.10.78. The *Shir 2's* development was watched attentively by the Ministry of Defence and by the MBT80 development team, and may have been an unofficial fall-back if MBT80 failed at an early stage. Document DOAE Note Number 134/200 in the National Archives DEFE 48/1076 file dated 1979 indicates that at this point there was serious consideration at some levels of buying 279 *Challengers* (referred to as such) – while the MBT80 programme proceeded (with in-service dates in the late 1980s) in order to bolster the Chieftain in BAOR. The MBT80 was officially cancelled in June 1980. Also see OD (79) 6 Cabinet Defence and Oversea Policy

Committee Sale of Tanks to Jordan, a Cabinet memo from the Defence Ministry dated 6 June 1979. After the cancellation of the Iranian FV4030 contracts, efforts were made at the highest levels to resolve the crisis caused by the cancellation. This memo indicates that the loss of the 1200 vehicle *Shir 2* order essentially wiped out 5 to 10 years' worth of steady work for ROF Leeds which required a minimum of fifty MBTs worth of annual orders to retain its skilled workforce. The document sets out the need to keep ROF Leeds open with foreign contracts of at least that many tanks per annum for 5 years to tide ROF Leeds over until MBT80 production could begin. This was a desperate attempt to keep the mechanism of production for the MBT80 in being without further British orders. It also gives some insight into the government's plight, which resulted in the eventual choice of the less expensive and timelier option in March of 1980 of buying the FV4030/4 – by then largely ready for production. The long term consequence was the privatization of Royal Ordnance's tank manufacturing plant and its eventual absorption by Vickers.

7. The decision to operate multiple types of MBT was also widely practiced in other NATO armies at the time and cannot be viewed with too much disparagement: the United States Army operated the M48A5, M60 series and the M1 Abrams in the early 1980s, the *Bundeswehr* operated the M48G1, the Leopard 1 series and the Leopard 2A4, and the Netherlands *Koninklijke Landmacht* operated the Leopard 1 and Leopard 2 (to name a few major armies). This was symptomatic of limited budgets for costly second generation MBT acquisition in all of the western democracies' armies. For the Ministry of Defence, the main consequence was problems caused by the small spares budget of the Challenger 1. This phenomenon is evident in the REME history nearly everywhere that the Challenger 1 is mentioned. See Kneen, Brigadier J.M. and Sutton, Brigadier D.J. OBE. *Craftsmen of the Army Volume II. The Story of the Royal Electrical and Mechanical Engineers 1969–1992*. Leo Cooper, Pen and Sword. Barnsley, UK. 1996.

8. See Griffin, Robert. *Challenger 1 Volume 1*. Kagero Publishing, Poland. 2013. The FV 4030/4 Challenger 1 purchase was approved because the vehicle was expected to be brought into service quickly and because much of its development was expected to have been completed. The Challenger turret included a proven mechanical stabilizer system, based closely on that employed on the Chieftain, with mechanical linkages. The IFCS fire control computer was also common to the late model Chieftains and considerably slower than the types being introduced in the early 1980s.

9. See Griffin, Robert. *Challenger 1 Volume 1*. Kagero Publishing, Poland. 2013. The Challenger 1 was provided with thermal imaging equipment (Pilkington Optronics TOGS) – although it lacked a comparable hunter killer FCS. Royal Ordnance was unable to sell the Challenger 1 despite its best efforts. Many of the Challenger 1's teething problems in 1985–86 related to the TN37 transmission but these were compounded by the scarcity of spares in BAOR.

10. The Leopard 2, from its earliest versions, was a vehicle characterized by simple design principles and sound engineering. Compared to the Challenger 1's spares related issues and the Chieftain's embarrassing availability rates it looked even more reliable in the eyes of many British soldiers. Some former soldiers who served at the time called this the 'Audi factor'.

11. The use of the US gas turbine with Allison transmission was one option seriously considered and evaluated for MBT80 and all British tank designs since 1980 have specified V type diesels. See National Archives File DEFE 24/1369. Vehicles. *Chieftain replacement and its Armament*. 26.09.78–31.10.78.

12. The West German Leopard 1 mounted the standard British RO 105mm L7 NATO gun used on most NATO tanks produced in the 1960s. The Soviets were quickly made aware of its performance during the Arab Israeli wars and of the limitations of the armour on their own tanks. When the T72 appeared in the Syrian arsenal in 1982 it was found that although Israeli 105mm L7 APDS rounds could punch through the glacis plate, they could not defeat the turret's frontal armour. This led the Russians to produce another version of the T72 with heavier hull armour which challenged the ability of many NATO MBTs to defeat its armour. The British L11 series of 120mm guns which armed the Chieftain and Challenger 1 were developed from 1961 and were adequate for this task; West Germany had, however, embraced the smooth bore concept in the early 1960s and, through early experiments in 90mm and 105mm, they eventually produced the Rheinmetall L44 smooth bore 120mm weapon. Production began in 1974, with the first version of the gun (known as the L/44 for its 44 calibre length) being tested on the preseries Leopard 2. The 120mm (4.7in) gun has a length of 5.28 metres (17.3ft.), and the gun system weighs approximately 3,317 kilograms (7,313lb). The 120mm smoothbore has since been developed extensively; a consequence of its widespread adoption and extensive development has allowed a wide range of extremely effective munitions to be developed for it. The HESH round for the British rifled L30A1 has not changed since the L11A5 gun was phased out in 1999 and it is now the rifled 120mm that suffers from a limited range of ammunition types.

13. The Rheinmetall Smoothbore (120mm L44), GIAT CN 120 G1, and L11A5 guns had comparable antitank performances firing kinetic energy rounds. The 47 ton Vickers Valiant was first designed in 1977 but not constructed in definitive form until the early 1980s. It was proposed unsuccessfully to several Middle Eastern

armies. Vickers had been given permission to produce Chobham armoured vehicles from 1977 onwards. The Vickers MBT Mk.4 Valiant basically employed a modernized Vickers MBT hull built in aluminium armour plate and mounting a steel turret with Chobham armour on the hull front and sides and on the turret front and sides. Its fire control system was the Marconi Centaur 1 and the turret was the original design template for that developed for the Vickers Mk.7 and thus the direct ancestor of the Challenger 2's turret. See p.211–213. Foss, Christopher F. and McKenzie, Peter. *The Vickers Tanks: from Landships to Challenger 2*. Patrick Stevens, UK, 1988.

14. See Griffin, Robert. *Challenger 1 Volume 2*. Kagero Publishing, Poland. 2013. The Royal Ordnance Challenger 1 was also demonstrated to the Egyptian Army in England in December 1985 without success. Royal Ordnance was privatized in January 1985 as Royal Ordnance PLC. In March 1987, after just over two years of operations as a publicly owned company, the ROF Leeds and the Challenger Tank programme's patents and intellectual rights were sold to Vickers. This effectively concentrated all British MBT manufacturing capability in one (private sector) company.

15. See p.228–232. Foss, Christopher F. and McKenzie, Peter. *The Vickers Tanks: from Landships to Challenger 2*. Patrick Stevens, UK, 1988. The Vickers Mk.7 MBT is an often ignored vehicle that was extremely promising in its time. Only a single prototype was constructed. The vehicle is also described in Dunstan, Simon. *Challenger 2 Main Battle Tank 1987–2006*. Osprey New Vanguard. UK, 2006.

16. See Dunstan, Simon. *Challenger 2 Main Battle Tank 1987–2006*. Osprey New Vanguard. UK, 2006. The Vickers purchase of ROF Leeds effectively concentrated all MBT manufacture in the country in one company with two manufacturing sites (in Leeds and Newcastle). The privatization of the Royal Ordnance tank factory was the first step in the consolidation of the armoured vehicle industry in the UK. Smaller AFV producers such as Alvis and GKN Sankey (GKN is an abbreviation for Guest, Keen and Nettlefolds) remained as separate military vehicle manufacturers operating individually for a few years but all were absorbed into Alvis-Vickers after the Cold War ended. Many of the smaller AFV producers had subcontracted turret design and manufacture, or manufacture of sub-assemblies, to Vickers prior to the mergers.

17. The M1s of 1st Platoon D Company 4th Battalion, 8th Cavalry Regiment won the CAT 87 competition, the *Bundeswehr's 124 Panzer Abteilung's* Leopard 2A4s took second place and 1st Platoon A Company 3rd Battalion, 64th Armoured Regiment, scored third in the competition. The M1 Abrams was triumphant at CAT87, and the American victory in the competition was followed by victory in the *Bundeswehr's Boselager* Cup by 1st Squadron 11th Armoured Cavalry Regiment, United States Army. American prestige had never been higher and reflected tremendous financial investment in advanced weapons programmes made during the Reagan administration. At this time the M1A1 was entering service in the US Army with the M256 120mm smoothbore, a licence-built version of the 120mm L44 Rheinmetall gun, although the CAT87 M1 Abrams mounted the US M68, a licence-built British L7 105mm rifled gun.

18. See Griffin, Robert. *Challenger 1 Volume 2*. Kagero Publishing, Poland. 2013.

19. See Griffin, Robert. *Challenger 1 Volume 2*. Kagero Publishing, Poland. 2013.

20. See Griffin, Robert. *Challenger 1 Volume 2*. Kagero Publishing, Poland. 2013. The Royal Hussars and the 2nd RTR both had their own 3 tank troop trained for the event, but because of operational commitments only 1 British team could participate, and The Royal Hussars were selected. The suitability of the Challenger 1 at this early stage in its career for the CAT 87 contest was doubtful; also due to very poor reliability of some of the elements of the fire control system, notably the Fire Control Computer Interface No.6, which was so prone to mefailure that REME had to institute a major programme to rectify deficiencies. Other major turret systems, like the TOGS night vision equipment, had to be retrofitted after delivery. See p.82. Kneen, Brigadier J.M. and Sutton, Brigadier D.J. OBE. *Craftsmen of the Army Volume II. The Story of the Royal Electrical and Mechanical Engineers 1969–1992*. Leo Cooper, Pen and Sword. Barnsley, UK. 1996. Naturally these facts were kept as discreet as possible in the hope of selling the Challenger 1 in the 1980s.

21. The CAT 87 disaster, its origins and the competition as lived by the team from The Royal Hussars is well described in Dunstan, Simon. *Challenger Main Battle Tank 1982–1997*. Osprey New Vanguard. UK. 1998. The scale of the embarrassment to the British Army of the Challenger 1's performance at CAT 87 is reflected by the fact that in CAT89 the British Army did not enter a team. A second account with an excellent selection of photos of The Royal Hussars in their first year as a Challenger regiment is available in Griffin, Robert. *Challenger 1 Volume 2*. Kagero Publishing, Poland. 2013

22. This was the first time that a British private company proposed a tank design to the British Army without at least the option of another equivalent or similar British design since the 1920s. The last time that Vickers had successfully managed to sell the British government a tank design 'off the shelf' was the Valentine in 1940 – and even that was a design equivalent to the A12 Matilda already in service!

23. Any Vickers proposal for a new MBT carried considerable political weight as it was the *last* British company capable of conducting mass production of a main battle tank.

24. The Vickers proof of principle contract was announced in parliament on 20 December 1988 by the Secretary of State for Defence, George Younger.
25. To give an idea of the development path followed with the Royal Ordnance L11A2, L11A3, L11A5, L11A6 (the modernized L11A3 fitted for the muzzle reference system) and L30A1 120mm guns, the original L11A1 gun L15 armour piercing discarding sabot (APDS) round fielded in 1965 could penetrate 340mm of RHA armour at 2000 metres, the L15A4 APDS-T round of the early 1970s could defeat 450mm RHA at 2000 metres, the L23 tungsten cored APFSDS round of 1983 could penetrate 450mm RHA at 2000 metres with greater accuracy. The CHARM1 L26A1 APFSDS-T round was the first to employ depleted uranium (DU) as the penetrator material due to its extreme density, permitting an armour penetration of 530mm at 2000 metres. The CHARM 3 L27A1 APFSDS-T round, also incorporating a depleted uranium penetrator, has a greater length to diameter ratio and can penetrate approximately 720mm RHA at this same range. These figures are estimates and not official figures. The barrel life on the L11 series gun was approximately half of that of the L30A1.
26. This was not the final chapter in the 140mm MBT armament developmental path, the next stage of which coincided with the Challenger 2's own definition as a production vehicle. A NATO project pursued between 1992 and 1998 by the UK, Germany, France and the USA put a 140mm gun into serious consideration as a standard armament for NATO's future MBTs. The programme was known by the acronym 'FTMA' for Future Tank Main Armament. As part of this project a wide variety of armament technologies were considered but a 140mm smoothbore gun firing one-piece ammunition was quickly defined by consensus as the short term solution. The United States built a 140mm armed demonstrator representing a possible configuration for the Abrams Block III MBT but discarded it due to funding cuts. The other three participant nations set up an equally divided commercial entity split between Royal Ordnance, Giat Industries and Rheinmetall Waffe Munition in 1997 to develop 140mm ammunition. In 1996 GIAT built a prototype turret, designated simply T4, to evaluate the feasibility of integrating the 140mm gun into the existing Leclerc design. Elsewhere German tests on the basis of the Leopard 2A4 and tests conducted by the Swiss RUAG firm for a 140mm gunned Leopard 2A4 were conducted. British contingencies are as yet unknown but would have certainly implicated the Challenger 2. The programme faded into obscurity after 1998.
27. See Dunstan, Simon. *Challenger 2 Main Battle Tank 1987–2006*. Osprey New Vanguard. UK, 2006. The original Rheinmetall 120mm smoothbore round (DM13) introduced in 1979 with the L44 gun had an armour penetration of 390mm of RHA at 2000 metres, followed by the DM23 of 1983 with a penetration of 470mm of RHA at 2,000 metres. These figures are estimates and are not official figures. The L/55, it was based on the same internal geometry as the L/44 and installed in the same breech and mount. The L/55 is 1.3 metres (4.3 ft.) longer, giving an increased muzzle velocity to ammunition fired through it. As the L/55 retains the same barrel geometry, it can fire the same ammunition as the L/44. One visual feature of the L55 is its length, the extra length provides longer range and it is suggested that the long barrel of the L55 has given it an increase of 1,500m; the muzzle velocities are also higher with the new barrel with APFSDS reaching 1,800mps.
28. See Dunstan, Simon. *Challenger 2 Main Battle Tank 1987–2006*. Osprey New Vanguard. UK, 2006. The rifled gun allowed the gun to fire spin-stabilized rounds with many of the same advantages as long range artillery. The L31 120mm HESH round can penetrate over 450mm of RHA, regardless of range, but is also a very effective weapon against infantry positions due to its blast effect. Fired at a relatively low velocity, this round has been favoured since 1967 as a direct fire, curved trajectory or anti-infantry round and has been tested extensively in combat in 1991 and 2003. HESH rounds could be fired at targets as far as 4,000–5,000 metres away with a good chance of a hit. The 7.62mm Chain Gun was standardized as the L94.
29. The Vickers Contract to build the Challenger recovery vehicle was awarded for thirty vehicles in 1985. These were later increased to seventy-four vehicles, and four were also later sold to Oman. See Foss, Christopher F. and McKenzie, Peter. *The Vickers Tanks: from Landships to Challenger 2*. Patrick Stevens, UK, 1988.
30. See Foss, Christopher F. and McKenzie, Peter. *The Vickers Tanks: from Landships to Challenger 2*. Patrick Stevens, UK, 1988. The number of changes made to the turret from Mk.7 to Challenger 2 was substantial and all the electronics were proofed in a special suppression chamber built at Vickers' Newcastle facility. The chamber was big enough for a prototype to drive in and traverse the turret 360 degrees to test under realistic conditions. The deciding element with all of the Challenger 2's systems and sub-systems was reliability. One of the features that had to be incorporated (which caused extensive discussion and evaluation) was the placement of the gunner's unity x1 window – which permitted a narrow forward viewing episcope for the gunner, but which had to be carefully placed in order to see around the TOGS barbette. Some of the changes and substitutions made during the evolution from Mk.7 turret to the Challenger 2 turret were as minor, others were much more significant. All were compromises between what the Army wanted and Vickers' initial designs.
31. See Foss, Christopher F. and McKenzie, Peter. *The Vickers Tanks: from Landships to Challenger 2*. Patrick Stevens, UK, 1988. The VS580-10 sight was first mounted on the Vickers Valiant. The Challenger 2 has a hull and turret built of rolled homogenous armour. While this work does not seek to discuss Chobham armour types at length

because the Challenger 2 is an in-service vehicle, Dorchester armour is a composite armour type that provides considerable immunity to kinetic and chemical energy rounds of all types. It is fixed in permanently mounted packs to the glacis and turret front of the Challenger 2 beneath a skin of steel armour plate. It can be fitted as appliqué kits as required to the vehicle's sides, lower front plate, or anywhere else deemed necessary to create multiple layer protection. Protection can also be supplemented with reactive armour or bar armour depending on the weight allowance for an operation. This same approach to protecting British MBTs to varying enemy threats by adding appliqué Chobham armour arrays to a base (frontally Chobham protected) vehicle was planned for the MBT80 in the 1970s. The protection philosophy has proved very successful in combat since 1991.

32. See Dunstan, Simon. *Challenger Main Battle Tank 1982–1997*. Osprey New Vanguard. UK. 1998. Also see Dunstan, Simon. *Challenger 2 Main Battle Tank 1987–2006*. Osprey New Vanguard. UK, 2006.
33. The decision to send the 7th Armoured Brigade (and subsequently the 4th Mechanized Brigade) to participate in the liberation of Kuwait as part of the American-led coalition was made because it was completely equipped with the two most modern AFVs in service with the British Army; the Warrior MICV and the Challenger Mk.2. The diminished Soviet threat had resulted in apparent neglect of spares and consumables for the MBT fleet in BAOR; this might have caused serious difficulties had troops been mobilized quickly to meet an emergency. See p.229–231 in Kneen, Brigadier J.M. and Sutton, Brigadier D.J. OBE. *Craftsmen of the Army Volume II. The Story of the Royal Electrical and Mechanical Engineers 1969–1992*. Leo Cooper, Pen and Sword. Barnsley, UK. 1996.
34. The measures necessary to round up enough Challenger spares to keep the armoured regiments in the 1st Armoured Division at 100% availability were by any description extreme. Between September 1990 and the following April, when the troops returned, the entire Challenger fleet in BAOR was immobilized and stripped for spares All Challenger assemblies harvested in this manner were checked over, packed and sent to Saudi Arabia by overworked teams of Royal Electrical and Mechanical Engineers. This also forced a re-evaluation of BAOR spares levels for the Challenger as its BAOR availability in mid-1990 had at times reached levels as low as 21%. See p.213, p.218–219 and p.229–231 in Kneen, Brigadier J.M. and Sutton, Brigadier D.J. OBE. *Craftsmen of the Army Volume II. The Story of the Royal Electrical and Mechanical Engineers 1969–1992*. Leo Cooper, Pen and Sword. Barnsley, UK. 1996. This is also described in Griffin, Robert. *Challenger 1 Volume 2*. Kagero Publishing, Poland. 2013, and in Dunstan, Simon. *Challenger Main Battle Tank 1982–1997*. Osprey New Vanguard. UK. 1998.
35. Vickers Defence in particular was heavily committed to supporting Operation Granby, as were most other defence industry manufacturers. See Foss, Christopher F. and McKenzie, Peter. *The Vickers Tanks: from Landships to Challenger 2*. Patrick Stevens, UK, 1988. For a precis of the Leopard 2's early career see Jerchel, M. and Schnellbacher, U., *Leopard 2 Main Battle Tank 1979–1998*. Osprey New Vanguard. UK. 1998. For a concise record of the early career of the M1 Abrams series see Zaloga, S. *M1 Abrams Main Battle Tank 1982–1992*. Osprey New Vanguard. UK. 1993. A more recent treatment of the Abrams's career in the period since Desert Storm is found in Green, M. *M1 Abrams at War*. Pen and Sword, Barnsley. 2005. The Leclerc is described in detail (in English) in Robinson, M.P. and Guillemain, Thierry. *Char Leclerc*. Kagero Publishing, Poland. 2015. While the British and West Germans were developing their own 120 mm guns, the French designed their own weapon for their new Leclerc MBT. The GIAT CN120–26 120mm smoothbore cannon, adopted as the CN120 F1, was a 52 calibre long gun. This weapon is capable of firing the same NATO standard 120mm rounds as the Leopard 2 and Abrams. The gun is insulated with a rigid thermal sleeve and employs an automatic compressed-air fume extraction system instead of the bore evacuator found on other guns. The Leclerc turret was designed on the modular concept to fit the autoloader, which allows a high rate of fire. The autoloader holds 22 rounds of ready ammunition, the most common types are the armour piercing fin-stabilized discarding sabot (APFSDS) with a tungsten core and the high explosive anti-tank (HEAT) round. There are 18 other rounds available for reload in the hull front. A Leclerc can fire accurately while traveling at speed on targets 4,000 metres away.
36. The Leclerc in particular received less attention from the British evaluation teams. This was partly because it was a very new design – the first prototypes having only been built in 1989–1990. The Leopard 2 and M1A1 were better known to the RAC which may have favoured these mature designs as contenders. The officer in charge of the RAC evaluation team for the Leclerc had other responsibilities in the programme. He was involved in the development of the 120mm L30A1 for the Challenger 2 and other gunnery projects. Operation Granby particularly interfered with the Leclerc evaluation as it mostly took place in France. The Leclerc was a swift and powerfully armed vehicle but French design philosophy did not embrace Chobham style armour and relied heavily on speed and acceleration for part of its protection. This factor, along with the 3 man crew and use of an autoloader was at odds with the British vision of an MBT. See Robinson, M.P. and Guillemain, Thierry. *Char Leclerc*. Kagero Publishing, Poland. 2015. The Leopard 2A5, M1A2 and Challenger 2 had all met the GSR requirements. The British, however, never favoured the M1 because its gas turbine engine's heat signature and high fuel consumption outweighed an excellent power to weight ratio. The Army had evaluated (and rejected) the possibility of using the gas turbine as a possible engine for the MBT80 in 1977–78 and had first tested a gas

turbine in a converted Conqueror hull in the 1950s. The Leopard 2A5, despite its many merits, was rejected as the least heavily armoured of the three vehicles. The Challenger 2 contract issued to Vickers as SR (L) 4026 was decided in July 1991 and valued in 1995 at £2 billion. The main sub-contractors were GEC Marconi (for turret systems and fire controls) and Royal Ordnance (for main and secondary armaments).

37. The British evaluation teams were more familiar with the Leopard 2A4 which had been evaluated by the MVEE in the preceding years while the M1A1 had been evaluated by the ADTU at Bovington.
38. The authors express their thanks to Dennis Lunn for his expert help with this précis of the careers of the 9 prototypes.
39. See Dunstan, Simon. *Challenger 2 Main Battle Tank 1987–2006*. Osprey New Vanguard. UK, 2006.
40. See Dunstan, Simon. *Challenger 2 Main Battle Tank 1987–2006*. Osprey New Vanguard. UK, 2006. Marc Chassillan, a French defence engineer who headed the development of the GIAT Leclerc recovery vehicle and who has studied and written extensively on modern main battle tanks for the French magazine RAIDS, wrote a very useful evaluation of the Challenger 2 amongst many other comparable vehicles in service around the world in *Tome 1* of his work entitled *RAIDS Hors Serie No.3 Les Chars de Combat en Action. Histoire et Collections. Paris 2001*. He notes that the Challenger 2 was the first British MBT to feature a significant percentage of foreign sourced subsystems and represented a major step away from Britain's traditional self-reliance for sighting and electronic components
41. The Mk.7 turret shell was used in the development of the Challenger 2 turret to prototype stage. While the production fire control system was sub-contracted to GEC Marconi, the fire control computer for the production vehicle came from CDC. This was not as simple as simply installing the Abrams' fire control computer; the software for the Challenger 2 was painstakingly developed by GEC Marconi and Vickers. Because the system that these companies created was developed along modular lines this substitution was integrated quickly and effectively. The CDC Fire Control Computer has two 32 bit processors and the fire control system includes a military standard 1553B Databus.
42. Barr and Stroud became Pilkington Optronics and was bought by Thales in 2001. Many updated versions of the third generation MBTs developed since the end of the Cold War have incorporated thermal channels in the commander's panoramic sights to permit thermal vision in a 360 degree field, and it is expected that this will be fitted to the Challenger 2 as a replacement for the TOGS system. See *BAE Systems bid provides glimpse into the future of Britain battle tanks*. Szondy, David. www.gizmag.com June 7 2016.
43. This interlude, which was naturally described in the British press as a reliability scandal, took about a year to work through. See Dunstan, Simon. *Challenger 2 Main Battle Tank 1987–2006*. Osprey New Vanguard. UK, 2006.
44. See Dunstan, Simon. *Challenger 2 Main Battle Tank 1987–2006*. Osprey New Vanguard. UK, 2006.
45. See Dunstan, Simon. *Challenger 2 Main Battle Tank 1987–2006*. Osprey New Vanguard. UK, 2006. The Type 37 regiments that originally received the Challenger 2 as complete equipment were the Royal Scots Dragoon Guards (1998), The 2nd Royal Tank Regiment (1998), The Queen's Royal Lancers (1999), The Queen's Royal Hussars (1999), The King's Royal Hussars and The Royal Dragoon Guards (2000). The 1st Royal Tank Regiment was partially equipped with the Challenger 2.
46. See Dunstan, Simon. *Challenger 2 Main Battle Tank 1987–2006*. Osprey New Vanguard. UK, 2006.
47. See Dunstan, Simon. *Challenger 2 Main Battle Tank 1987–2006*. Osprey New Vanguard. UK, 2006. Also see Vickers Defence Systems, Press Release 'Vickers And UK MOD Sign Base Repair Contract For Challenger 2.' V.D.S. May 28 2002.
48. See Dunstan, Simon. *Challenger 2 Main Battle Tank 1987–2006*. Osprey New Vanguard. UK, 2006. The Type 58 regiment had approximately the same combat value as (and a similar tactical organization to) the Cold War era Type 57 Regiment, with 4 sabre squadrons.
49. For a full listing of problems encountered and a listing of Royal Armoured Corps and other British Army units involved in Saif Sareea see p.22–25, and p.44. Comptroller and Auditor General's Report, *Ministry of Defence: Exercise Saif Sareea II* (HC 1097, Session 2001–02). HMSO, UK. 2001. Both this document and the HC502 report attacked the Ministry of Defence's approach to planning and conducting the exercise. See *Ministry of Defence: Saif Sareea II*. House of Commons Committee of Public Accounts, (HC 502. Sixth Report of Sessions 2002–2003), HMSO, UK. March 2003.
50. 'In planning future exercises, the Department should balance the cost of modifying key equipment against that of supporting unmodified equipment in theatre. In this case, the decision on grounds of economy not to "desertise" the Challenger 2 Main Battle Tank increased the costs and decreased the effectiveness of the Exercise.' p.8 *Ministry of Defence: Saif Sareea II*. House of Commons Committee of Public Accounts, (HC 502. Sixth Report of Sessions 2002–2003), HMSO, UK. March 2003. The cost overrun on the exercise totalled over £26 million. On p.9 of the same document the fact that dust had already been pointed out as a cause of problems during exercises at BATUS prior to *Saif Sareea II*. In the Minutes of Evidence section of the same report Lieutenant-General Sir John Reith CB, CBE answered committee questions regarding the subsequent

dust ingestion testing of the Challenger 2 with side appliqué armour arrays and dust skirts in Canada. An availability figure of 83% was given for Challenger 2 during operations in Kosovo in comparison with about 30% during *Saif Sareea II*.

51. p.7 *Ministry of Defence: Saif Sareea II*. House of Commons Committee of Public Accounts, (HC 502. Sixth Report of Sessions 2002–2003), HMSO, UK. March 2003.
52. The cost ramifications of the exercise were of great concern to the government and to the army. See p.42 and 49. Comptroller and Auditor General's Report, *Ministry of Defence: Exercise Saif Sareea II* (HC 1097, Session 2001–02). HMSO, UK. 2001. The ADTU subsequently tested improved components to address the hot air recirculation and sand filter issues highlighted in the exercise. See Dunstan, Simon. *Challenger 2 Main Battle Tank 1987–2006*. Osprey New Vanguard. UK, 2006.
53. See p.6 item 1 of House of Commons Committee of Public Accounts Report. *Ministry of Defence. Operation Telic – United Kingdom Military Operations in Iraq.* 39th Report of Session 2003–2004 (HC273) HMSO, UK. 16 September 2004. The value of Exercise Saif Sareea in preparing for Operation Telic was confirmed in this report, especially in terms of logistics support essential to deploying an armoured division. The deployment took approximately half as long as the Operation Granby deployment phase in late 1990.
54. See p.25 Ev. 2 of House of Commons Committee of Public Accounts Report. *Ministry of Defence. Operation Telic – United Kingdom Military Operations in Iraq.* 39th Report of Session 2003–2004 (HC273) HMSO, UK. 16 September 2004. While 1st Armoured Division was expected to have to be capable of fighting Iraqi armour in the first stage of the campaign, the constituent brigades of the 1st Armoured Division were deliberately configured with as much infantry as possible due to the fact that urban combat was anticipated in the Basra area. The British plan had originally included an advance into Iraq through Turkey and into northern Iraq, but was changed to an advance on Basra through the Al-Faw peninsula. This breakdown of the forces in the 7th Armoured Brigade is explained in *The Eagle and Carbine 2003–2004. The Regimental Magazine and Association Report*. The Royal Scots Dragoon Guards. Edinburgh, 2004. Details of the battlegroup operations are also set out on p.14 of the *Army Board of Inquiry Report into the Circumstances Surrounding the Deaths of 24848863 Cpl Steve Allbutt QRL and 25119984 Tpr David Clark QRL…at Al Basra Iraq March 25th 2003*.
55. See p.40 Ev. 18 of House of Commons Committee of Public Accounts Report. *Ministry of Defence. Operation Telic- United Kingdom Military Operations in Iraq.* 39th Report of Session 2003–2004 (HC273) HMSO, UK. 16 September 2004. The uparmouring kits were not delivered until 20 March 2003. See *The Eagle and Carbine 2003–2004. The Regimental Magazine and Association Report*. The Royal Scots Dragoon Guards. Edinburgh, 2004 for an account of the SCOTS DG's preparations for the advance into Iraq.
56. See p.7 item 7, and see p. 11 item 15 of House of Commons Committee of Public Accounts Report. *Ministry of Defence. Operation Telic – United Kingdom Military Operations in Iraq.* 39th Report of Session 2003–2004 (HC273) HMSO, UK. 16 September 2004. While the lessons of Exercise Saif Sareea II had been successfully implemented (on the mechanical side of operational requirements) by the time that Operation Telic was launched, supply scales were still in some respects inadequate. One particularly disturbing feature of the Challenger 2's deployment in Operation Telic was that operational NBC filters for the tanks NBC systems were not available in time for deployment and had not even been delivered by July 2003, which risked the need to operate these vehicles with the crews wearing NBC suits had any chemical weapons been encountered. This would have caused much difficulty had enemy armour simultaneously been encountered *en masse*. See p.28 Ev. 6, ibid. Lieutenant-General Robert Fry's statements indicate that nearly all of the other AFV types operated by 7th Armoured Brigade had their filters issued prior to commencing the operation. Also see Beaton, Maj. A. (Ed.) *The Fight for Iraq, January–June 2003: The British Army's Role in Liberating a Nation*. Army Benevolent Fund, U.K. 2004.
57. See Dunstan, Simon. *Challenger 2 Main Battle Tank 1987–2006*. Osprey New Vanguard. UK, 2006. Dunstan's work recounts this and several other incidents in the first phase of Operation Telic in excellent detail.
58. See p.8–12. *Army Board of Inquiry Report into the Circumstances Surrounding the Deaths of 24848863 Cpl Steve Allbutt QRL and 25119984 Tpr David Clark QRL…at Al Basra Iraq March 25th 2003*. It is apparent from reading this report that extensive precautions were taken to provide adequate Identification Friend or Foe (IFF) measures and that the TOGS equipment in the vehicles involved in the incident were in excellent working order. The incident was caused by a confused handover and confusion over unit boundaries amongst other causes.
59. See p.8–12. *Army Board of Inquiry Report into the Circumstances Surrounding the Deaths of 24848863 Cpl Steve Allbutt QRL and 25119984 Tpr David Clark QRL…at Al Basra Iraq March 25th 2003*.
60. See *The Eagle and Carbine 2003–2004. The Regimental Magazine and Association Report*. The Royal Scots Dragoon Guards, Edinburgh, 2004 for an account of A Squadron, SCOTS DG in this engagement. Naturally the presence of embedded journalists has given much cause for political discussions but the main cause for mention here is the extreme sensitivity this caused to casualties amongst the British forces engaged in the operation.
61. See Dunstan, Simon. *Challenger 2 Main Battle Tank 1987–2006*. Osprey New Vanguard. UK, 2006. Dunstan's work recounts this and several other incidents in the first phase of Operation Telic in excellent detail.

62. See *The Eagle and Carbine 2003–2004. The Regimental Magazine and Association Report*. The Royal Scots Dragoon Guards, Edinburgh, 2004, for an account of C Squadron, SCOTS DG in this engagement.
63. See *The Eagle and Carbine 2003–2004. The Regimental Magazine and Association Report*. The Royal Scots Dragoon Guards. Edinburgh' 2004, for an account of C Squadron, SCOTS DG in this detached engagement. The incident is also recounted in Dunstan's book and his account is based on interviews with the crew members. See Dunstan, Simon. *Challenger 2 Main Battle Tank 1987–2006*. Osprey New Vanguard. UK, 2006.
64. See *The Eagle and Carbine 2003–2004. The Regimental Magazine and Association Report*. The Royal Scots Dragoon Guards. Edinburgh, 2004, for an account of B Squadron, SCOTS DG in this engagement.
65. The front ROMOR armour array was composed of the steel frames and brackets that supported Royal Ordnance developed explosive reactive armour bricks known as ROMOR A. ROMOR A is an explosive reactive armour type that sandwiches a charge composed reputedly of DEMEX 200 explosive between two steel plates held out from the face of steel armour by the frames and brackets. When impacted by a kinetic energy or chemical energy projectile, the ROMOR A panel detonates, effectively detonating chemical rounds prematurely or disrupting kinetic energy rounds. The sets employed on the Challenger 1 were manufactured in Nottingham and were procured in in early 1991 prior to operation Granby for use on the Challenger 1 MBTs lower front plates and on Centurion AVRE turrets. Equipment from these same sets was later employed on the Challenger 2 in 1999 and subsequently in 2003–2006, for the protection of the lower glacis (which was not then protected with Chobham armour). The side arrays carried by the Challenger 1 during Operation Granby were recycled as well, and were improved with dust skirts and fitted to the Challenger 2. These were designed by Vickers and were manufactured by Royal Ordnance and Vickers in 1990–1991. See Foss, Christopher F. and McKenzie, Peter. *The Vickers Tanks: from Landships to Challenger 2*. Patrick Stevens, UK, 1988.
66. See House of Commons Committee of Public Accounts Report. *Ministry of Defence. Operation Telic – United Kingdom Military Operations in Iraq*. 39th Report of Session 2003–2004 (HC273) HMSO, UK. 16 September 2004. A variety of factors are listed in the report as causes for Iraqi disaffection with the occupying forces, including the unwillingness of the coalition commander to order the army to impose order immediately following the defeat of Saddam's army. Some details of the SCOTS DG's departure and the handover are recorded in *The Eagle and Carbine 2003–2004. The Regimental Magazine and Association Report*. The Royal Scots Dragoon Guards. Edinburgh, 2004.
67. The use of 'bar' or slat armour has been employed extensively as an effective anti-shaped charge or chemical energy round measure since the beginning of the 1980s. It is particularly effective against RPGs and the most recent phenomenon in the race between the shaped charge and the tank, pursued with great ingenuity since the Panzerfaust was first employed in 1944. The use of bar armour is cost effective and relatively light but is easily damaged and can affect the crew's movement when entering and exiting the vehicle.
68. The Army did not announce the incident for some months while they implemented the armour improvements described. The ROMOR A reactive armour installation of the Challenger 1 and Challenger 2 lower hull front was designed in 1990 (or possibly before) to protect the lower front plate against single warhead chemical energy attack (RPG, antitank guided missile, or HEAT rounds) and proved vulnerable to tandem warhead shaped charge attack such as the RPG29. This type of armour installation provides little protection against IED (improvised explosive device) attack which are usually remotely detonated bombs made from buried medium artillery shells set off when a tank drives over them. A related threat is the EFP or explosively formed projectile, which has an outer dish shaped liner over a conically formed shaped charge, which becomes a concentrated projectile when detonated. Both IED and EFP type weapons can be buried under roads or set up to detonate into the flank of a passing vehicle. They are most dangerous to the underside of an MBT.
69. These measures to improve the Challenger 2's armour were made public in 2007–2008.
70. The TELEX RWS was not fitted to all of the Basra tanks but is an option available for TES equipment if needed. The GMG is an automatic grenade launcher. During 'Operation Charge of the Knights' Sergeant C.P. Richards, RDG was awarded the Military Cross for carrying the advance through 5 IED ambushes in his up-armoured Challenger 2 at the head of the British advance back into Basra.
71. The 116 Challenger 2s deployed in Operation Telic in 2003 were made up of elements of three different armoured regiments each of which left part of their strength at their home bases. When the British left Basra City on 2 September 2007 to consolidate the army base at Shaibah Challenger 2s from the lone sixteen tank squadron escorted the convoy. The British Army did not leave Iraq until March 2011.
72. These engineers' vehicles are normally operated in a brigade level detachment in squadron strength, which is usually broken into troops assigned to different battle groups. Obstacle crossing is of paramount importance on the modern battlefield as is minefield clearance.
73. The MTU diesel still remains an excellent potential upgrade for the British Challenger 2 fleet because it offers a reserve of power that would support further upgrades (and weight gain) and permit cohesion with several other NATO armies using the Leopard 2 series. The existing CV12 has, however, proven a very reliable engine and the current maintenance programme achieves excellent availability rates.

74. Since the end of the Cold War, the Challenger 2 has lost numerous bids for sales to foreign armies. In contrast, the Leopard 2 has been adopted in Sweden, Spain, Denmark, Norway, Finland, Greece, Poland and Canada to name but a few NATO users. Krauss-Maffei has displayed tremendous initiative and resourcefulness in buying up surplus Leopard 2s and selling them in South America and Asia.
75. See *SDPM 0001305 238 Minutes of the Special AASB (international Offers) Held at 9H30 08 July 1998 at DHQ. Pretoria. (Declassified 2014.03.23)*. The South African trials held in June 1998 came down to the Challenger 2E winning out narrowly versus the more complex Leclerc EAU, the difference being the more generous Vickers financing and limited co-production stance for a proposed order of 108 tanks. The order was never placed due to changes in South Africa's government; Project *Aorta* remains unfulfilled and the South Africans have yet to buy a new MBT. Despite the slight preference overall of the South African Army for the Challenger 2 evident in ARMSCOR documents dated 13 May 1998 declassified in 2014, reservations were noted regarding the 120mm L30A1 because only the UK and Oman had adopted the weapon – presumably because of the possible impact on ammunition cost and availability. This issue very probably also affected the evaluation in the Hellenic Army MBT selection trial 5 years later.
76. Both BAe and General Dynamics UK are focused on lighter AFV designs and have retreated significantly from the capability of building a complete MBT at time of writing. Both companies have placed bids to modernise the Challenger 2 fleet in company of other domestic and foreign companies. With so many Challenger 2s mothballed or dismantled, it is shameful that no funds have been made available to examine reconfiguration of these vehicles into a comprehensively modernised MBT to rectify the existing tank's compatibility with fixed 120mm ammunition or to mount a new turret of British design. Furthermore the potential need for specialised armoured vehicles of the heavy APC type, or for tanks optimised for an urban combat environment with an alternate main armament capable of high elevation, should not be ignored. Regrettably it is likely that most of the Challenger 2s presently out of service – over 100 vehicles – will end up being scrapped (a process may indeed be underway). It is especially wasteful to scrap such a costly and potentially long-lived weapons platform when other countries can rebuild and reinvent AFVs based on substantially less capable platforms.
77. p.8 *The Strategic Defence and Security Review and the National Security Strategy; Sixth Report of Session 2010–12*. (HC 761). House of Commons Defence Committee. HMSO. 2011. UK.
78. The King's Royal Hussars based at Tidworth is actually a Type 44 Regiment with only 3 MBT squadrons. The long periods where regiments like the Kings Royal Hussars and The Royal Tank Regiment were substantially deployed to Afghanistan left the RAC at its lowest ebb in terms of available MBT units since the 1920s as recounted in detail in the regimental journals for both regiments up to 2013. In that period the Royal Wessex Yeomanry was used to complete trainable squadrons in both units with reservists at a time when the budget could barely support paying Territorial armoured crewmen. See *The Kings Royal Hussars Association Newsletters* for the period 2010–2014 and *Chain Mail Magazine* Royal Wessex Yeomanry 2014.
79. The 3rd Division has since (late 2016) been cut to 2 Strike Brigades although the 3 MBT regiments have been retained. The most recent MBT regiments to lose their Challenger 2s, the Royal Dragoon Guards and the Royal Scots Dragoon Guards, both faced a choice of amalgamation or changing into armoured cavalry regiments.
80. The totals have been quoted in slightly different terms but reliable sources seem to estimate that less than 250 vehicles will be retained. It is sad to recognize that the remainder will probably be scrapped when the Germans have done such fantastic work rebuilding and reinventing their stock of stored Leopard 2 variants. See *Janes Defence Review*, C.F. Foss. 2015. '… *in mid-2013, the MoD confirmed that it was working on a Challenger 2 Life Extension Programme (LEP) as the name suggests a more austere project, with less ambitious scope than the CSP that would standardize the remaining vehicles in a more cohesive configuration. This coincided with a significant reduction in strength, with the number of available vehicles now reduced to 227, with 58 nominally issued to each of the three Royal Armoured Corps regiments.*'
81. The use of depleted uranium in ammunition has become a political question in several countries.
82. The production of rifled 120mm ammunition in Belgium has since been examined, although details are lacking. As Rheinmetall now offers a 130mm gun the 120mm L55 may soon be superseded and replaced by a common autoloaded turret on NATO's MBTs… though that eventuality is purely conjecture.
83. The fleet management programme is more properly known as Joint Asset Management and Engineering Solutions (JAMES) – essentially a system that pools the Challenger 2 fleet. Excepting a single squadron per regiment issued as training vehicles, and a full regiment's worth of operationally ready vehicles, roughly half of the Army's operational vehicles are kept in storage. These exclude all vehicles undergoing repair or rebuild but the programme serves to minimise costs.